Love's Madness

Love's Madness

Medicine, the Novel, and Female Insanity
1800–1865

HELEN SMALL

CLARENDON PRESS · OXFORD

1996

Oxford University Press, Walton Street, Oxford OX2 6DP

Oxford New York
Athens Auckland Bangkok Bombay
Calcutta Cape Town Dar es Salaam Delhi
Florence Hong Kong Istanbul Karachi
Kuala Lumpur Modras Madrid Melbourne
Mexico City Nairobi Paris Singapore
Taipei Tokyo Toronto
and associated companies in
Berlin Ibadan

Oxford is a trade mark of Oxford University Press

Published in the United States
by Oxford University Press Inc., New York

British Library Cataloguing in Publication Data
Data available

Library of Congress Cataloging in Publication Data
Small, Helen.
Love's madness: medicine, the novel, and female insanity,
1800–1865 / Helen Small.
Includes bibliographical references (p.   ).
1. English fiction—19th century—History and criticism.
2. Literature and mental illness—Great Britain—History—19th century.
3. Women and literature—Great Britain—History—19th century.
4. Mentally ill women in literature. 5. Loss (Psychology) in literature.
6. Medicine in literature. 7. Love in literature. I. Title.
PR868.M46S63 1995 823'.809353–dc20 95-33313
ISBN 0-19-812273-X

1 3 5 7 9 10 8 6 4 2

Typeset by Best-set Typesetter Ltd., Hong Kong
Printed in Great Britain
on acid-free paper by
Bookcraft Ltd.
Midsomer Norton, Avon

In loving memory of my parents,
Wenda and Colin Small

Preface

> The fashion of such narratives . . . changes like other earthly things.
>
> (Sir Walter Scott, *The Fortunes of Nigel*, 1822)

LOVE'S MADNESS is about the history of a convention. When Sir Walter Scott remarked on the changeability of narrative fashions in 1822, he was referring specifically to the contemporary taste for lavish descriptions of weddings. *Love's Madness* is concerned with a less happy theme, but one which fascinated Scott's contemporaries—and Scott himself—almost as much. Stories about women who go mad when they lose their lovers were extraordinarily popular during the late eighteenth and early nineteenth centuries, attracting novelists, poets, dramatists, musicians, painters, and sculptors. The subject was by no means new, as any reader of Shakespeare knows, but with the cult of sensibility it took on an unprecedented importance. The representative figure of madness ceased to be the madman in chains and became instead the woman whose insanity was an extension of her female condition.

In *Abandoned Women and the Poetic Tradition* (1988), Lawrence Lipking claims that poetry is the only literary medium capable of conveying the intensity of the love-mad woman's desolation. The figure of the abandoned woman stands, in nearly every culture, 'for all the inexpressible yearnings that poetry alone can begin to approach'. Because she makes us feel the futility of the words which cannot bring back an unfaithful or a dead lover, we become aware of 'what language leaves out. A lover of language needs this reminder; without a sense of unfulfillment, lovers might grow too fat.' The abandoned woman must therefore scorn 'the contentment of prose'. Plot is alien to her, traumatized by the loss of her lover. Unwilling or unable to cease lamenting, she arrests the narrative movement on which prose fiction has for the most part depended.[1]

---

[1] Lawrence Lipking, *Abandoned Women and the Poetic Tradition* (Chicago, 1988), 28, 30.

Lipking's assertion that the medium of prose cannot do justice to the subject of female abandonment is misleading, but it may help to explain the circumstances under which love-madness first became a popular subject for fiction. With its disregard for narrative momentum, its intense focus on the isolated moment of pathos, and its alertness to the inadequacy of language (the characteristic dashes, ellipses, and heavy marks of emphasis; the fracturing of narratives under the pressure of feeling), the eighteenth-century sentimental novel brought to the subject of love-madness many of the qualities which Lipking sees only in poetic treatments of the theme. Novelistic interest in the love-mad woman persisted, however, after the demise of sensibility. Far from taking their woes back to poetry when the cult of sensibility declined, love-mad women continued to inspire large numbers of prose works (as well as writing in other genres) throughout the next century. Why, when the cultural conditions which sustained the sentimental vogue had so largely changed, and when the vogue for female insanity was increasingly singled out for attack by anti-sentimentalists, did love-mad women prove so durable?

Love's Madness traces the fortunes of love-mad women in fiction and in medicine between about 1800 and 1865. These dates are not the usual dividing lines for English history, but they make sense in this instance. In literary terms, they demarcate the period between the decline of sentimentalism and the emergence of sensation fiction (the genre most commonly associated with female insanity in current criticism of the Victorian novel). In medical terms, they mark out a key stage in the history of insanity, beginning with major reform initiatives at the turn of the century and ending with the establishment in 1865 of the Medico-Psychological Association (later the Royal College of Psychiatrists). Much of *Love's Madness* is concerned with what it means to refer to literature and medicine as separate fields. The attack on sensibility helped to change a climate in which medicine and fiction had shared much the same language and the same assumptions. By the 1860s relations between them had become far more complex.

Despite the wider culture's continuing interest in love-mad women, the convention's tendency to deflect or suspend plot severely limited its usefulness for doctors seeking to develop a therapeutic approach to insanity. The Ophelian type becomes harder to find in medical texts as the nineteenth century progresses, and when she does appear her meaning is very carefully negotiated. Love-madness gives way to a view of femininity in which female disturbance is no longer externalized in the narrative of desertion, but internalized into the very workings of the body. However, the same factors which made the love-mad woman an intractable subject for medical writers made her a figure of peculiarly enduring interest to writers of fiction. Her intense claim on the reader's sympathy, her traumatized state, and not least her very conventionality made her remarkably well adapted to serve as a vehicle for focusing other, less readily articulated kinds of complaint. In the verse tradition, the abandoned or love-mad woman typically tells her own story. In the novel, she is almost always talked about: observed, described, and sympathized with by a stranger. By placing love-mad women within other, more complex narratives, nineteenth-century narrative fiction increasingly made new demands of the convention, teasing out the ways in which its implications might change when viewed in relation to different social, political, and historical concerns. In the process, the nineteenth-century novel changed the convention and changed itself.

Many people have helped me in researching and writing this book. I wish first and foremost to thank Gillian Beer, who has been unfailingly generous with her time and knowledge. I am grateful also to Dorothea Barrett, who helped me considerably in the first stages of my research, and to Daniel Pick and Rod Mengham for their careful reading of an early version of *Love's Madness*. My understanding of the medical history of this period has been assisted by Simon Schaffer, who read Chapter 2 and made helpful criticisms of it; and by Alison Winter, who brought an acute historian's eye to bear on Chapters 1 to 5. I am indebted to Roy Porter, and to the editors and readers of *History of the Human Sciences*, for detailed responses to Chapter 2, a version of which appeared in that journal in 1994.[2] George Donaldson, Heather Glen, John Lyon, and Sheila Stern commented valuably on specific chapters. I owe particular thanks to Glen Cavaliero, who read *Love's Madness* with meticulous care; and to Stephen Wall, whose

---

[2] Helen Small, ' "In the Guise of Science": Literature and the Rhetoric of Nineteenth-Century English Psychiatry', *History of the Human Sciences*, 7 (1994), 27–55.

advice has made this a better book. John Kerrigan's criticisms were almost as valuable as his encouragement.

Thanks are due also to Wyn Beasley for drawing my attention to the Sarah Fletcher memorial at Dorchester Abbey, Dorchester-on-Thames, and providing me with an initial transcription of the text; to the Reverend John Crowe, Rector of Dorchester Abbey, who was helpful in providing unpublished material relating to Sarah Fletcher; and to Richard Luckett, who allowed me to benefit from his knowledge of the early nineteenth-century Handel repertoire. Lord David Cobbold kindly gave me permission to use the library and archives of Knebworth House. I am grateful to Clare Fleck for her assistance at Knebworth, and to Marie Mulvey Roberts and John Sutherland for sharing their work on Rosina Lytton.

St Catharine's College, Cambridge appointed me to a Research Fellowship for three years from 1990, giving me the opportunity to take my Ph.D. research further and see it to near-completion as a book. The Department of English, University of Bristol, gave me special research leave in order to see the book into press. I am grateful also to the following for assistance at various stages: Stefan Collini, Clare Conolly, Kate Flint, Peter Garside, Simon Gaunt, Linda Hardy, Paul Hartle, Mark Kilfoyle, Jonathan Lacey, Nigel Leask, D. F. McKenzie, Ruth Morse, Adrian Poole, Suzanne Raitt, James Raven, Fiona Russell, Naomi Tadmor, and Trudi Tate.

H.S.

March 1995

Contents

Illustrations

I

Love's Madness

Reader!
If thou hast a Heart fam'd for
Tenderness and Pity,
Contemplate this Spot.
In which are deposited the Remains
of a Young Lady, whose artless Beauty,
innocence of Mind, and gentle Manners,
once obtain'd her the Love and
Esteem of all who knew her. But when
Nerves were too delicately spun to
bear the rude Shakes and Jostlings
which we meet with in this transitory
World, Nature gave way; she sunk
and died a Martyr to Excessive
Sensibility.

M*rs.* SARAH FLETCHER
Wife of Captain FLETCHER,
departed this Life at the Village
of Clifton, on the 7 of June 1799,
In the 29 Year of her Age.
May her Soul meet that Peace in
Heaven, which this Earth denied her.

SARAH FLETCHER's gravestone is set in the floor of Dorchester Abbey at Dorchester-on-Thames. As the record of an actual and fatal case of nervous breakdown, it is a reminder that what was a subject for fiction was also a matter of life and death. Eighteenth-century lapidary inscription was often self-displayingly literary, but in this case the similarity to a narrative convention then very much in vogue is particularly marked. The story the gravestone tells is the staple material of its period's novels, and many of its features—the opening appeal to the sensitivity of the reader, the outline of the virtues of the heroine, the adoption of the language of sensibility—

reinforce the remarkably close parallel between this public record of a woman's death and the language of popular fiction.

Sentimentalism establishes this tombstone's claim on the visitor's emotions, yet it is remarkable how little of Sarah Fletcher's life is actually revealed. Only the brief outlines of a tragic decline and early death are given, so that the tale gains some of its poignancy from the very shadowiness with which it is sketched. An emblematic spareness of plot and characterization allows the epitaph to refer beyond itself, recalling the deaths of numerous fictional heroines equally artless in their beauty, innocent in their minds, gentle in their manners, and fragile in their nerves. The gravestone uses the language of sentimentalism to identify a type of nervous frailty and to identify also a community of readers who will respond to that frailty with the appropriate sensibility. Sarah Fletcher herself is absent from the presentation of her case. The specific origins and symptoms of her sickness are elided, depersonalized, and rendered as nature's response to the common trials 'we meet with' in this life, so that her death finally emerges as a signal but abstract illustration of excessive sensitivity—a martyrdom, no less. The conflicting ethics of the contemporary cult of feeling are clearly on display in that awkward panegyric 'a Martyr to Excessive Sensibility'.[1]

The appeal to the conventions of sensibility had a specific purpose in this instance, for Sarah Fletcher's death was far less commonplace in its manner than the distanced belletrism of the inscription might suggest. This was an epitaph genuinely concerned to establish for itself *only* those 'gentle readers' whose hearts were 'fam'd for Tenderness and Pity'. On Saturday 15 June 1799 *Jackson's Oxford Journal* had reported Sarah Fletcher's demise in these terms:

On Saturday last, an Inquest was taken at Clifton, in this county, before R. Buckland, Gent. one of his Majesty's Coroners, on the body of Mrs. Sarah Fletcher. This lady put an end to her existence by hanging herself with her pocket handkerchief, which she fastened to a piece of small cord, and affixed it to the curtain rod of the bedstead in the room in which she usually slept. After a full investigation of the previous conduct of the deceased, and the derangement of her mind appearing very evident from the testimony of the gentleman at whose house this unfortunate affair

---

[1] See John Mullan, *Sentiment and Sociability: The Language of Feeling in the Eighteenth Century* (Oxford, 1988), 201–40, on the strains within the late 18th-cent. cult of sensibility.

happened, as well as from many other circumstances, the Jury, without hesitation, found a Verdict—*Lunacy*. The husband of this unfortunate lady is an officer in the Navy, and is now on his passage to the East Indies.[2]

Jackson's Journal appeared to provide the particulars missing from the gravestone, but it continued to conceal a great deal by its vague gesture towards 'many other circumstances'. Though the full story of how and why Sarah Fletcher came to be hanging from her own bedstead in June 1799 was never fully ascertained, even at the coroner's inquest,[3] it was, from the day of her death, a subject of popular discussion and dispute in Dorchester and its environs, as it continues to be. In 1989 an Abingdon bank clerk was walking by the Thames at Dorchester, when he claims to have met the ghost of Mrs Fletcher.[4] This is only the most recent of numerous reported sightings over the last 200 years.

The most extensive and accessible account of Sarah Fletcher's death is contained in a pamphlet published for Dorchester Abbey, written by a visitor to the church in 1913 who became fascinated by the memorial and by reports of the ghost. Mrs Ffoulkes's 'Story of Sarah Fletcher' was pieced together principally from the recollections of Mr Poyntz, Vicar of Dorchester at the time of her visit, and from the testimony of a previous vicar who had lived in the Fletchers' house as a child. Though it could still accommodate an explanation of the suicide in terms of nervous collapse or 'derangement', this version declines to use the vocabulary of sickness or insanity:

'In the last years of the eighteenth century,' said [Mr Poyntz], 'Sarah Fletcher (one of the Fletchers of Saltoun) lived at Clifton Hampden—not far from here. Captain Fletcher was in the Navy, and, following the popular tradition of the sea, he was not only inconstant, but unfaithful. He actually proposed marriage to a wealthy heiress living some distance away, and he was on the point of committing bigamy when Mrs Fletcher was warned at the last moment—she had only just time to reach the church to stop the ceremony.

---

[2] *Jackson's Oxford Journal*, 15 June 1799, 3.

[3] The coroner's inquest report on Sarah Fletcher's death is no longer in existence, but the contents are alluded to in an MS account of her life and death by Mrs Ffoulkes. The full text of Mrs Ffoulkes's research into Sarah Fletcher's death and into the sightings of her ghost was published as the main feature of a booklet entitled *Dorchester Ghost Stories* sold in aid of the Abbey Monastery Guest House, probably in the 1970s (n.d.). A shortened version, entitled *The Story of Sarah Fletcher*, is still on sale to visitors at the Abbey.

[4] Related to the current Rector of Dorchester Abbey.

'It is not difficult to imagine the scene which followed . . . Captain Fletcher literally ran away, made for London, and sailed for the East Indies. The unwedded bride returned home with her parents, and Sarah Fletcher drove back to Clifton Hampden and hanged herself in her bedroom fastening her pocket handkerchief to a piece of cord which she fixed to the curtain rod of her bedstead.'[5]

Where *Jackson's Oxford Journal* implied a history of mental illness, citing the 'previous conduct of the diseased' as a key factor in the jury's verdict of 'Lunacy', Mrs Ffoulkes suggests that, on the contrary, suicide was an immediate and lucid response to a devastating betrayal. With the insanity verdict, she indicates, the inquest jury simply avoided a difficult judgement on the motives which led a deserted wife to suicide—motives which were, after all, likely to seem self-evident and deserving of sympathy, although self-murder could not be openly condoned under the law. Equally to the point, a *non compos mentis* verdict obviated the need for a moral response. It removed the case from the category of felony, allowing the relatives to give the body a Christian burial, and preventing the forfeiture of property to the Crown.[6] At a time when the individual, social, and spiritual implications of self-murder were being heavily debated, the insanity verdict allowed some suspension of judgement, even as it implied a move away from the Church's jurisdiction towards the less finally condemnatory judgement of medicine.[7]

That there *was* considerable uncertainty about how to interpret Sarah Fletcher's death is underscored by a postscript to Mrs Ffoulkes's pamphlet, written by a local historian, Edith Stedman, probably in the late 1960s:

This is by no means the end of the thriller. Local legend has it that she was just buried at the crossroads with a stake through her heart as was the custom in those days for suicide. If that were true, how does [Sarah] Fletcher end up in the abbey with an elaborately eulogistic inscription in a very prominent place? I have seen the inquest which seemed very cagey. . . . (My own theory is that she was murdered by her husband or

---

[5] Ffoulkes, 'The Story of Sarah Fletcher', fol. 3.

[6] E. Umbreville, *Lex Coronatoria* (London, 1761), 310–11. The standard history of coroners' laws in England is Thomas Rogers Forbes, 'Crowner's Quest', *Transactions of the American Philosophical Society*, 68/1 (1978). See pp. 36–9 on suicides. Also, Nigel Walker, *Crime and Insanity in England*, 2 vols. (Edinburgh, 1968), i. 136.

[7] Michael MacDonald and Terence R. Murphy, *Sleepless Souls: Suicide in Early Modern England* (Oxford, 1990).

through his connivance and she wants her name cleared of the sin of suicide.)[8]

Even without pursuing Stedman's speculations, it is clear that the gravestone's elaborately sentimental references to nervous illness serve a function beyond the ordinary requirements of mourning. Sarah Fletcher's interment in Dorchester Abbey seems to have been permitted only after some debate over whether she should have the solace of church burial (not until 1823 was an act passed to prevent the superstitious disposal of the bodies of suicides[9]). The public record of her death is remarkable in the end not for what it says about her but for what it finds itself unable to say. In its oblique references to nerves 'too delicately spun', 'excessive sensibility', nature giving way, the text of the tombstone disguises self-murder (or what was judged to have been self-murder) and conceals also considerable doubt about how to interpret it. Should this woman be held culpably responsible in taking her own life or, by labelling her suicide a product of mental derangement, should she be excused responsibility and spared the consequences of guilt? The specific appeal of Sarah Fletcher's memorial to a sentimental convention of nervous debility is the product of a set of uncertainties concerning her status in the Church, in law, in medicine, and in the eyes of her society, all of which suggestively underlie its closing sentence. Once her history is known, a note of pressing concern is brought to what might otherwise be a fairly standard appeal to the Mercy of God: 'May her Soul meet that Peace in Heaven, which this Earth denied her.' It is a benediction but it is also, given its context, a question.

The shadowy story of Sarah Fletcher's death may now seem merely sentimental, but for nineteenth-century visitors to Dorchester it would have had a stronger resonance. Tales of women driven to insanity and, in many cases, to suicide by the death or treachery of their lovers were more than just a free-standing literary convention. They were deeply ingrained in the culture's conception of femininity, as they had been for centuries. In its widest aspect, the literary type invoked by Sarah Fletcher's memorial can be traced as far back as Greek tragedy. English eighteenth-century fiction inherited the

---

[8] Edith G. Stedman, unpub. postscript to Mrs Ffoulkes's 'Story of Sarah Fletcher'. 'Sarah' is placed in square brackets here because Stedman mistakenly refers to her throughout as 'Lucy'.

[9] Forbes, 'Crowner's Quest', 38–9.

convention principally from an extensive complaint literature popularized in broadside ballads and poems over the previous centuries, and stemming back variously through chronicle complaint to Boccaccio, and through pastoral epic and epistolary laments to Theocritus, Lucian, Virgil, and Ovid.[10] This was one of the most eloquent faces of madness, testifying over and over again to women's truth and men's falsity, to the mental vulnerability of a woman deserted and the callousness of the man who could leave her.

Female complaint thus has a long tradition, but until the late seventeenth century love-madness was more likely to be depicted as a male malady than as a female one. Burton's *The Anatomy of Melancholy* (1621) devotes one-third of its space to the subject of 'Love Melancholy', tracing its history, its symptoms, and its reputed cures through classical and medieval literature to his own day. For Burton, unsatisfied love was, almost by definition, a form of sickness and its principal subject was the male tragic hero. Unlike many of his classical sources, Burton perceived love as a disease not of the body but of the mind, affecting 'both imagination and reason'.[11] The precise causes of the affliction were numerous—the stars, diet, country, clime, condition, idleness, dancing, kissing, and amorous tales could all bring on an attack—and the possibilities for cure were discouragingly limited: the lover might be persuaded out of his affliction, might remove himself from the cause, but unquestionably the most effective remedy was gratification. If a cure was not forthcoming, the prognosis for the love-melancholic was poor:

For such men ordinarily as are throughly possessed with this humor, become senseless and mad, *insensati & insani* . . . and as I have proved, no better than beasts, irrational, stupid, head-strong, void of feare of God or men, they frequently forsweare themselves, spend, steale, commit incests, rapes, adulteries, murders, depopulate Townes, Cities, Countries to satisfie their lust. . . . Hee that runnes head-long from the top of a rocke, is

[10] John Kerrigan (ed.), *Motives of Woe: Shakespeare and 'Female Complaint'. A Critical Anthology* (Oxford, 1991), 1. See also Lawrence Lipking, *Abandoned Women and the Poetic Tradition* (Chicago, 1988), 2; *Loving Mad Tom: Bedlamite Verses of the XVI and XVII Centuries*, with illustrations by Norman Lindsay, foreword by Robert Graves, text ed. Jack Lindsay, musical transcriptions by Peter Warlock (London, 1927); and, on the medieval background to this literature, Mary F. Wack, *Lovesickness in the Middle Ages: The 'Viaticum' and its Commentaries* (Philadelphia, 1990).

[11] Robert Burton, *The Anatomy of Melancholy*, 3 vols., ed. Thomas C. Faulkner *et al.*, with intro. by J. B. Bamborough (Oxford, 1989–94), iii. 58.

not in so bad a case, as hee that falls into this gulfe of Love. For . . . saith Gordonius the prognostication is, they will either runne mad, or dye. For if this passion continue, saith *Ælian Montaltus, it makes the blood hot, thicke, and blacke, and if the inflammation get into the braine, with continuall meditation and waking, it so dries it up, that madnesse followes, or else they make away themselves* . . . Go to Bedlam for examples. It is so well knowne in every village, how many have either died for love or voluntary made away themselves, that I need not much labor to prove it . . .[12]

Despite the appeal to the experience of 'every village', Burton's dense allusions to classical and medieval literary example are proof that he comprehended love-melancholy almost exclusively within the framework of a masculine tradition of heroic action (indeed, the adjective 'heroical' as used in *The Anatomy of Melancholy* derives from an ancient term for love-melancholy passed down from the earliest Greek physicians, through Arab and Roman medicine, into the medieval and Renaissance literature on love-madness[13]). For readers a century later, however, the form in which love-madness was most likely to be contemplated had substantially changed. The feminization of madness is often taken to be a phenomenon of the late eighteenth and early nineteenth centuries,[14] but in the case of love-madness that gender shift occurs nearly 100 years earlier. The precise reasons for the change are not clear. What can be said is that the gradual decline during the seventeenth century in the prominence of male love-melancholy in favour of its (already well-established) female counterpart paved the way for the eighteenth century's primary emphasis on women as the typical sufferers from debility of the nerves.[15]

One version of the love-madness convention, roughly contemporaneous with Burton's *Anatomy*, proved more powerful than any other in symbolizing and to some extent assisting this change in the gendering of love-madness in English culture. Shakespeare's *Hamlet*

[12] Burton, *Anatomy of Melancholy*, iii. 197–9.
[13] John Livingstone Lowes, 'The Loveres Maladye of Hereos', *Modern Philology*, 11 (1913–14), 491–546. On Elizabethan attitudes to 'the lover's malady', see also Lawrence Babb, *The Elizabethan Malady: A Study of Melancholia in English Literature from 1580 to 1642* (East Lansing, Mich., 1951), 128–74.
[14] See particularly Elaine Showalter, *The Female Malady: Women, Madness, and English Culture 1830–1980* (London, 1987), 8.
[15] On the gender balance of reported cases of insanity in this period (and more broadly), see Michael MacDonald, *Mystical Bedlam: Madness, Anxiety, and Healing in Seventeenth-Century England* (Cambridge, 1981), 35–40.

offers two contrasting images of love-madness: one deeply ambiguous, the other classically 'authentic'. Hamlet's behaviour persuades
some of those around him to believe that when Ophelia obeys her
father and denies the Prince her company, his sanity is endangered
as a result. She, at least, registers his appearance when he comes to
her closet as that of a stereotypical love-melancholic:

>   his doublet all unbrac'd
> No hat upon his head, his stockins fouled,
> Ungart'red, and down-gyved to his ankle,
> Pale as his shirt, his knees knocking each other,
> And with a look so piteous in purport . . .   (II. i. 75–9)

Polonius, charged with discovering the cause of Hamlet's distem-
per, initially grasps at his daughter's explanation and attempts to
persuade Claudius and the Queen that Hamlet has indeed fallen
through dispriz'd love

> into a sadness, then into a fast,
> Thence to a watch, thence into a weakness,
> Thence to a lightness, and by this declension,
> Into the madness wherein now he raves.   (II. ii. 147–50)

Claudius' scepticism is probably closer to an audience's likely response, faced with the riddling, often inconclusive evidence of
Hamlet's mental condition. Ophelia, by contrast, presents a painfully direct instance of the lover's malady. Although Shakespeare
remained faithful to the established terms of the love-madness
convention, his 'document in madness'[16] quickly supplanted all
earlier madwomen in the English dramatic repertoire. Ophelia set a
standard for later madwomen. Decked with 'fantastical garlands',
her hair loose, playing upon a lute, singing bawdy ballads and sad
laments, she proved a compelling theatrical figure. That she remained so was perhaps due to the fact that she so successfully
combined 'dual messages about femininity and insanity': at once a
wistful innocent and a sexually explicit woman.[17] Ophelia was, as
Elaine Showalter has demonstrated, supremely manipulable—the
most famous love-mad woman of all would be fashioned and
refashioned to meet the tastes of different eras and different audiences, sometimes the bawdy wench, sometimes the sentimental
love-melancholic.[18]

---

[16] Laertes describing Ophelia, *Hamlet*, IV. v. 178.
[17] Showalter, *The Female Malady*, 10–11, 90–4.
[18] Elaine Showalter, 'Representing Ophelia: Women, Madness, and the Responsi-

Ophelia had a Shakespearean sister who, though less written about, is, in some ways, even more revealing. *The Two Noble Kinsmen* (*c*.1613), a collaborative work between Shakespeare and Fletcher, represents both a tribute to *Hamlet* and, in certain significant respects, a reworking of it. In a scene probably by Fletcher, 'The Wooer' tells 'The Gaoler' that his daughter has been found, 'distracted' by her unrequited love of Palamon:

> The place
> Was knee-deep where she sat; her careless tresses
> A wreath of bulrush rounded; about her stuck
> Thousand fresh water-flowers of several colors,
> That methought she appeard like the fair nymph
> That feeds the lake with waters, or as Iris,
> Newly dropp'd from heaven. Rings she made
> Of rushes that grew by, and to 'em spoke
> The prettiest posies—'Thus our true love's tied,'
> 'This you may loose, not me,' and many a one ... (IV. i. 82–91)

The story of the Gaoler's Daughter replays that of Ophelia, drawing upon the same iconography and much the same conception of female psychology. Like Ophelia's ravings, her words are essentially lucid in content, harping continually on the unkindness of her lover and pathetically imagining alternative stories to her own. Here, as in *Hamlet*, insanity discloses a surprising degree of sexual knowledge, but it also lends a curious innocence to the bawdiness, an innocence emphasized in the Wooer's description ('as Iris | Newly dropp'd from heaven'). Both Ophelia and the Gaoler's Daughter are traumatized at a moment of disappointment in love which is inextricably intertwined in their understanding with the death of their fathers (actual in *Hamlet*, imagined in *The Two Noble Kinsmen*):

> She sung much, but no sense, only I heard her
> Repeat this often—'Palamon is gone,
> Is gone to th' wood to gather mulberries.
> I'll find him out tomorrow.'
> ... Then she talk'd of you, sir—
> That you must lose your head to-morrow morning,
> And she must gather flowers to bury you,
> And see the house made handsome. Then she sung
> Nothing but 'Willow, willow, willow', and between

bilities of Feminist Criticism', in Patricia Parker and Geoffrey Hartman (eds.), *Shakespeare and the Question of Theory* (New York, 1985).

Ever was 'Palamon, fair Palamon,'
And 'Palamon was a tall young man'.   (IV. i. 66–82)

In all this, *The Two Noble Kinsmen* remains fairly close to *Hamlet*. Unlike the earlier play, however, this reworking of love-madness provides a doctor for its sufferer. The Wooer reports that when the Gaoler's Daughter saw herself spied upon, she 'straight sought the flood', but this Ophelia is not permitted to drown. Saved by the Wooer she is returned to her father and brother, and they seek the advice of a physician. Act IV, scene iii, in which the Doctor first sees her in the full force of her 'green sickness', was probably Shakespeare's work, and it functions as a hinge between tragedy and comedy in the play. 'How prettily she's amiss!', the Doctor marvels, while the Gaoler's Daughter raves on about love; 'how her brain coins!' When asked by the Gaoler 'What think you of her, sir?', the Doctor's answer initially sounds like an admission of professional defeat: 'I think she has a perturb'd mind, which I cannot minister to.' In fact, he means that the maid's senses can only be restored by the man whose failure to love her caused her madness—or by a man willing to 'take upon [him] . . . the name of Palamon'. And there's the rub. Sexual satisfaction is here the cure for the maid's sexual derangement, but the cure operates with a cursoriness which is evidently comic at her expense. Despite the assurance that only the man who drove her mad can bring her back to sanity, in practice it appears that any man might satisfy her in Palamon's name. Moreover, whereas derangement made the Daughter a pretty coiner of language, cure leaves her subdued, far less articulate, and deferential to the lover who has brought her sexual fulfilment.

In most of its details, the story of the Gaoler's Daughter anticipates the convention of female insanity later handed down to nineteenth-century writers. The comic resolution of *The Two Noble Kinsmen*, however, is atypical of the direction the literary figure of the love-mad woman would take. By the eighteenth century, few such women would be allowed the restoration of their senses promised by Shakespeare and Fletcher's doctor, and enacted more fully in William Davenant's reworking of the play later in the century.[19] They become instead isolated, iconic figures whose condition is so quintessentially feminine as to make treatment redundant.

[19] Davenant's *The Rivals* was first performed in 1662.

While the Ophelian type remained central in the history of English representations of female insanity from the seventeenth century onward, its enduring appeal cannot be fully understood without a recognition of the other ways in which the convention was kept both alive and topical. An extensive broadsheet literature on female insanity continued to circulate. Mad 'Tom of Bedlam' had his counterparts in lovesick 'Bess of Bedlam' and 'Mad Maudlin' from at least the beginning of the seventeenth century,[20] and, even in the nineteenth, the street ballad and theatre song tradition to which they primarily belonged often proved the testing-ground for more lasting poetic endeavours. During the Renaissance and Restoration periods, the mad song was frequently the musical or dramatic show-piece of a play, and an actress could make her name with it. 'The Mad Shepherdess' famously secured the fortunes of Mrs Davis. Hearing her sing the touching lament for a lost lover in a performance of *The Rivals*, Charles II was 'so pleased, that he took her off the stage and had a daughter by her'.[21] Few stories better demonstrate the degree to which irrationality had become part of the dramatic repertoire of feminine sexuality. Moreover, the mad-women of street ballads and theatre songs can offer pointed comparisons with the Ophelias of their day. During the Augustan period, for example, when Ophelia was in one of her most sober and decorous phases, the Restoration theatre-goer was able to hear in Anne Bracegirdle's performances of mad songs by John Eccles a less restrained version of mania arising from love. His 'I burn, my brain consumes to ashes'[22] and 'Oh! Take him gently from the pile . . . and I will scorch for him the while' are two of the more unabashed items in an extensive Restoration repertoire, including works by Henry and Daniel Purcell, John Weldon, Godfrey Finger, and John Blow.[23]

At the turn of the nineteenth century, when Sarah Fletcher met her tragic end, the convention was still very much alive. Between about 1770 and about 1810, stories about bereaved or deserted

---

[20] Lindsay (ed.), *Loving Mad Tom*, 13, 63.

[21] Recorded in Thomas Evans (ed.), *Old Ballads, Historical and Narrative, with some of Modern Date* (London, 1784), 285. Also *The Dramatic Works of Sir William D'Avenant*, 5 vols. (Edinburgh, 1872–4), v. 217–20.

[22] On the association between madness and burning, see Babb, *The Elizabethan Malady*, 132–3.

[23] Catherine Bott has recorded 18 Restoration mad songs (CD no. 433 187-2 Decca–L'Oiseau Lyre; London, 1993).

women fallen into insanity were the subject of an extraordinary
vogue in sentimental prose, poetry, drama, and painting.[24] For a
culture which placed great emphasis on the visible expression of
sentiment and on the dramatic staging of 'feeling', the established
iconography of love-madness was attractive material. Three versions of the story stand out as having particularly influenced other
writers and artists of the period. Henry Mackenzie's *The Man of
Feeling* (1771) almost single-handedly transformed the asylum
from a grotesque Augustan theatre of irrationality into an arena for
sentimentalism.[25] The madwoman encountered by Harley on his
visit to Bedlam leads him—and led many a willing reader with
him—to tears. Forbidden by her father to marry the poor man she
loves, she has been driven insane by news of the young man's death
from a fever in the West Indies and by her father's remorseless
insistence that she marry 'a rich miserly fellow . . . old enough to be
her grandfather'.[26] Trauma induces a hypnotic repetition of her
symptoms in this, the most pathetic and attractive inmate of the
asylum: 'My Billy is no more!', she laments: '. . . do you weep
for my Billy? Blessings on your tears! I would weep too, but my
brain is dry; and it burns, it burns, it burns!'[27] When Harley weeps
to see the madwoman's pain, their tearful dialogue threatens to
break down the distinction between the man of feeling and the
woman deranged by feeling. Nevertheless, the difference is re-
asserted by a pointed recollection of his economic power over her.
Harley and his friend pay the asylum-keeper for their encounter
with the tender spectacle of distressed female sensibility and in so
doing effectively buy their way out of the threat of seeming like her.
As Philip Martin notes, 'At this moment the madwoman is replaced
in her original position as commodity, sold for the public's gaze, as
indeed she was to be sold again to the readers of Mackenzie's novel,

---

[24] Robert Mayo, 'The Contemporaneity of the Lyrical Ballads', *PMLA*, 69
(1954), 486–522, lists 39 poems with mad and deserted women. See also Philip
Martin, *Mad Women in Romantic Writing* (Brighton, 1987), 19–27.
[25] The asylum was not a new subject for literature. It was a favourite preoccupation of 18th-cent. fiction, though typically depicted as a place of anarchic violence
rather than a forum for sentimentality. See Max Byrd, *Visits to Bedlam* (Columbia,
SC, 1974); and Byrd, 'The Madhouse, the Whorehouse and the Convent', *Partisan
Review*, 44 (1977), 268–78.
[26] Henry Mackenzie, *The Man of Feeling*, ed. with intro. by Brian Vickers
(Oxford, 1967), 33. Mackenzie's madwoman is strongly indebted to Davenant's
Celania in *The Rivals*.
[27] Mackenzie, *Man of Feeling*, 34.

who purchased access to the promotion of feeling, the exercise of the truly warm and sympathetic sensibility.'[28] Much the same commercialization of the spectacle of feminine derangement is evident in the reception of Lawrence Sterne's love-mad Maria, in *Tristram Shandy* (1760–1)[29] and *A Sentimental Journey* (1768),[30] though Sterne's version of the theme was more provocatively erotic than Mackenzie's. When Yorick weeps for Maria, she takes his sodden handkerchief from him and places it in her bosom to dry. Maria had inspired more than thirty paintings by the start of Victoria's reign, most successfully Angelica Kauffman's engraving of *c.*1777, which circulated all over Europe in print form, and was transferred to numerous fashion articles, including a watch-case, a tea-waiter, and a Wedgwood cameo.[31]

A third version of the sentimental love-madness convention was probably the most generative of all. William Cowper's depiction of Crazy Kate in *The Task* (1785), like Sterne's Maria, attracted numerous painters (notably Fuseli and Shepheard), and a flood of imitative poetic and prose narratives in the magazines and miscellanies of the 1790s.[32] The gently deranged servant girl, pathetically obsessed with the lover who deserted her, was still prompting close imitations well into the nineteenth century. Matthew 'Monk' Lewis's poem 'Crazy Jane' was in circulation in street ballad form for several years before it was published in his small 1812 collection of verse. Although Lewis's version proved to be one of the less lasting products of the sentimental tradition, its influence was enormous, inspiring an aggressively unsentimental series of recastings by Yeats as late as the 1930s.[33] In its own day, Monk Lewis's ballad gave rise to a parlour song, a play (C. A. Somerset's *Crazy Jane*, performed at the Surrey Theatre, Drury Lane in 1824), and a hat. With Crazy Jane millinery, female insanity reached its nadir, be-

[28] Martin, *Mad Women in Romantic Writing*, 19.

[29] Laurence Sterne, *The Life and Opinions of Tristram Shandy*, ed. Graham Petrie, with intro. by Christopher Ricks (Harmondsworth, 1967), vol. ix, ch. 24 (pp. 600–2).

[30] Laurence Sterne, *A Sentimental Journey through France and Italy by Mr.* ed. Ian Jack (London, 1968), 113–16.

[31] Catherine M. Gordon, *British Paintings of Subjects from the English Novel 1740–1870* (New York, 1988), 73–6, 79, 87–9.

[32] Martin, *Mad Women in Romantic Writing*, 19.

[33] See *W. B. Yeats: The Poems*, ed. Richard J. Finneran, corr. repr. (London, 1984), 255–60, 582.

coming a fashion accessory: madness quite literally *à la mode*. The
poem's closing stanza can stand as a representative expression of
the degree to which the sentimental madwoman had become a
virtually self-sustaining literary type, inoculated against the harsher
resonances of social reality even as it invited a sensitive reading:

> Gladly that young heart received him,
>   Which has never loved but one!
> He seemed true, and I believed him;
>   He was false, and I undone.
> Since that hour, has reason never
>   Held her empire in my brain:
> Henry fled: with him for ever
>   Fled the wits of Crazy Jane![34]

A concluding rhyme verging on insensitivity ('Jane' paired with
'brain') and a banally literal dramatization of the masculine hold on
reason are now more likely to provoke laughter than tears, and
many early nineteenth-century readers would have felt the same
way. By 1812 sentimentalism had long been under concerted attack. While the narrative of the love-mad woman had proved its
fundamental appeal many times over, it had become a convention
in need of revision.

When Sarah Fletcher hanged herself in her bedroom in 1799,
literary models were thus readily at hand for those who mourned
her. Medical theory of the day was equally well equipped to pronounce upon this quintessentially feminine condition. Between the
epitaph's allusions to delicate nerves and excessive sensibility, the
blunter reference to derangement in *Jackson's Oxford Journal*, and
the jury's verdict of lunacy, there lies a fluid mass of medical and
sub-medical theorizing. Contemporary physicians, faced with a
situation in which a woman hanged herself a few hours after her
husband had attempted bigamy and then deserted her, would have
had a range of explanations to draw upon. As Janet Oppenheim has
noted, a medical practitioner's diagnosis would have depended
largely upon his personal style. Nervous exhaustion, overwrought
sensibility, hysteria, melancholia, mania, or the more somatic but
vague 'gynaecological disturbance' were all possible choices.
'Erotomania' was less likely, but it is worth mentioning because it

---

[34] Matthew Lewis, *Poems* (London, 1812), 24–5: 25.

has attracted a certain amount of notice from historians. Though sometimes invoked in cases of suicide, or of 'jealous fury, producing sanguinary catastrophes', erotomania was, by the eighteenth century, a highly specific syndrome, classed by the President of the Royal College of Physicians in Edinburgh, Alexander Morison, as a subspecies of monomania. As the term indicates, it was a species of insanity arising from love, but the object of affection was not necessarily male, real, or even animate. Though, like 'the melancholy of disappointed love', it was usually met with in young females, Morison held that erotomania could be distinguished by the presence of delirium, and he warned against a loose confusion of terms.[35]

Whichever label the physician chose, he would have been drawing on a long medical tradition of viewing the physiology of women as cripplingly vulnerable to their emotional state. It is most likely that he would have opted for one of two standard diagnoses: melancholia (Morison's 'melancholy of disappointed love') or hysteria. Janet Oppenheim's description of the typical Victorian response to mental distress in women would already have held true in 1799:

Unlike male nervous breakdown, the element of personal choice or responsibility was rarely granted much influence in women. Some medical writers found it simplest to deal with female nervous collapse under the great catch-all classification of hysteria, still the archetypal feminine functional nervous disorder in the nineteenth century. While the more grotesque physical symptoms of hysteria—the full-scale paroxysm, temporary convulsions, palsies, motor and sensory impairments, respiratory obstructions, and speech disorders—bore no resemblance to the manifestations of depression, the less dramatic ones certainly did. Hysterical women might abstain from food, suffer sudden fits of weeping and experience chronic lassitude in a manner that was indistinguishable from profound melancholy.[36]

---

[35] Alexander Morison, *Outlines of Lectures on Mental Diseases*, 3rd edn. (London, 1826), 49. Erotomania is clearly distinguished from nymphomania, which is held to be a 'local irritation', whereas erotomania had its seat 'wholly in the mind' (p. 49). Morison was drawing heavily on the work of the renowned French physician Jean-Étienne Dominique. See his entry on erotomania in the *Dictionaire des sciences médicales, par une société de médecins et de chirurgiens*, 60 vols. (Paris, 1812–22), xiii. 186–92. Esquirol's case-studies were, however, notably more sentimental than his theory (e.g. pp. 189–90).

[36] Janet Oppenheim, *'Shattered Nerves': Doctors, Patients, and Depression in Victorian England* (Oxford, 1991), 181.

'In grave cases', particularly when the body had been weakened by pregnancy, childbirth, or simply by menstruation, 'suicide itself might prove the tragic finale in this drama of exhaustion.'[37]

Any nineteenth-century reader with a classical education, or who had had occasion to consult a dictionary on the subject, would have known that the word 'hysteria' derived from *hustera*, the Greek for 'uterus'. The 1797–1801 edition of the *Encyclopaedia Britannica* defined it as follows:

HYSTERIC AFFECTION, or *Passion*, (formed of υςερα [*sic*] 'womb'); a disease in women, called also *suffocation of the womb*, and vulgarly *fits of the mother*. It is a spasmodico-convulsive affection of the nervous system, proceeding from the womb . . .[38]

Ilza Veith's celebrated account of the illness *Hysteria: The History of a Disease* (1965) found medical references to the womb's pathological influence on female behaviour as early as c.1900 BC. Although more recent historians have disputed the hysterical content of the uterine disorders recorded in the Egyptian Kahun papyrus and early Hippocratic writings,[39] the origins of the *Encyclopaedia Britannica*'s definition clearly lie very far back. The typical hysterical symptom, still documented in many nineteenth-century descriptions of the disease, was the choking lump (*globus hystericus*) rising in the throat. It owed its aetiology largely to the second-century physician Galen, who held that hysteria arose from engorgement of the womb. By the start of the nineteenth century it had long been established that any influence the womb exerted must be indirect, but, although psychological influences had begun to be taken much more seriously, the belief in a link with the reproductive system remained crucial. It was not until the late nineteenth and early twentieth centuries that the clinical work of Jean-Martin Charcot, and, later, of Sigmund Freud and Joseph Breuer, made hysteria's psychological origins accepted throughout

---

[37] Oppenheim, *'Shattered Nerves'*, 189.

[38] *Encyclopaedia Britannica*, 3rd edn., 20 vols. (London, 1797–1801), vix. 49.

[39] Ilza Veith, *Hysteria: The History of a Disease* (Chicago, 1965), 1–2. Her interpretation of the papyri has been challenged by Harold Merskey and Paul Potter, 'The Womb Lay Still in Ancient Egypt', *British Journal of Psychiatry*, 154 (1989), 751–3. For the rebuttal of Veith's account of the early Hippocratic writings, see Helen King, 'From Parthenos to Gyne: The Dynamics of Category', doctoral dissertation, University College London, 1985, and 'Once upon a Text: The Hippocratic Origins of Hysteria', in George Rousseau and Roy Porter (eds.), *Hysteria in Western Civilization* (Los Angeles, forthcoming).

the medical profession; and though medical writers had always conceded that hysteria had its occasional male victims, it was not until the massive incidence of shell-shock in the First World War that hysteria ceased to be thought of as an essentially female condition.

One clinical feature of hysteria remains unchanged throughout its long history, and that is the extraordinary flexibility of its symptomatology. Hysteria is defined today as a psychological condition characterized by its ability to mimic physical diseases. Until the second half of the twentieth century (when it has been largely superseded by anorexia and pre-menstrual syndrome), it could claim to have been the feminine performative disease *par excellence*. It was, above all, a disease whose meaning was as culturally and historically specific as that of femininity itself.[40] Most discussion has concentrated on its role as a pathological by-product of Victorian gender roles, to be found especially in idealized young middle- and upper-class womanhood,[41] but hysteria was an established diagnostic category by the end of the eighteenth century, well recognized and routinely treated by physicians.[42] Almost all the sufferers were women (98 per cent, according to Guenter Risse's study of the eighteenth-century records of the Edinburgh Infirmary), most were unmarried, and most were in their twenties or thirties.[43] Only a minority of the patients in Risse's study displayed the characteristic fits. The traits ascribed to hysteria increased markedly in range and number during both the eighteenth and nineteenth centuries. By the end of that period the hysteric had *carte blanche* when it came to self-expression. She could possess any (and would very likely give signs of several) of the following characteristics: a nervous temperament, violent and unstable emotions, depression, excitement, poor attention span, disturbed intellect, disturbed will, deficient judgement, dependency, immaturity,

---

[40] Mark S. Micale, 'Hysteria and its Historiography: The Future Perspective', *History of Psychiatry*, 1 (1990), 32–124. See also Showalter, *The Female Malady*, *passim*; Stephen Heath, *The Sexual Fix* (Basingstoke, 1982), 27–49; and Barbara Ehrenreich and Deirdre English, *Complaints and Disorders: The Sexual Politics of Sickness* (New York, 1973), 19–48.

[41] Micale, 'Hysteria and its Historiography', 39, 86.

[42] Katherine E. Williams, 'Hysteria in Seventeenth-Century Case Records and Unpublished Manuscripts', *History of Psychiatry*, 1 (1990), 383–401.

[43] Guenter B. Risse, 'Hysteria at the Edinburgh Infirmary: The Construction and Treatment of a Disease 1770–1800', *Medical History*, 32 (1988), 1–22.

egocentricity, attention-seeking, deceitfulness, theatricality, simulation, jealousy, fearfulness, and irritability.[44]

The class stereotype of the hysteric holds out less well under scrutiny. Eighteenth- and nineteenth-century fiction and autobiography have undoubtedly helped support a tendency to think of hysteria as a luxury illness belonging to the leisured world of emotional intrigues, novel-reading, smelling-salts, and oppressive fathers. Yet novels also give indications of a larger picture. As any reader of Smollett or Fielding knows, eighteenth-century ladies'-maids were vulnerable too. In fact, hysteria was widespread among the working classes. All the female patients in Risse's study were from labouring backgrounds. Robert Brudenell Carter, writing his treatise *On the Pathology and Treatment of Hysteria* in 1853, noted that any visitor to Hanwell Asylum might see how prevalent hysteria was among working women, and devoted a final chapter to its frequency among the poor. Although only 25 years old when he wrote this much-cited volume, Carter was nothing if not worldly in his tone. He allowed himself several gibes about the number of hysterical attacks arising from fear 'among servant-girls, and in fourth-rate boarding schools',[45] and rebuked the vulgarity of working-class hysterics, eager to report on their experience of the speculum and to request its renewed application.[46] Though Carter's misogyny was not necessarily typical, his sense that hysteria had become *démodé* and *déclassé* was widely shared. Hysteria was, by 1853, a thoroughly familiar condition, often more indicative of malingering than of a fine sensibility. For various reasons, the once-fashionable sentimental stories about tender love-melancholics had become just as problematic for medicine as they had for fiction.

The response to Sarah Fletcher's death thus drew on a long literary tradition and at least as long a medical tradition, but what was the relationship between the two? Literary representations have had a significant role in shaping the history of female insanity, but in

---

[44] Chris N. Alam and H. Merskey, 'The Development of the Hysterical Personality', *History of Psychiatry*, 3 (1992), 135–65.

[45] Robert Brudenell Carter, *On the Pathology and Treatment of Hysteria* (London, 1853), 31.

[46] Ibid. 70. He continues: 'Such persons, a few years ago, were almost limited to the class whose hysteria depended directly upon the frustration of desire; but the recent increase of uterine disease has produced many of them, among the tertiary cases resulting from general discontent.'

approaching the representation of nervous illness in literature we need to recognize that (as in Sarah Fletcher's case) it was already the stuff of fiction in the real world. When we look at madness in literature we are looking at a representation of something that is *already* representation. That is, we look at an experience for which the language of description is inescapably representational, and whose pathos is muted because language has been overworked. Indeed, critical theory of the last three decades has hotly debated whether the ability to command pathos gives literature a claim to speak with exclusive authority for and about insanity. Most influentially in the work of Michel Foucault and Shoshana Felman, literature has been seen as possessing a privileged relationship to madness: a capacity to gesture beyond rationalism and beyond words towards the emotional tenor of an experience otherwise silenced by the language society gives us.[47] Sarah Fletcher's gravestone is particularly relevant here because it puts in historical perspective the emergence of such a sharp distinction between different claims to knowledge of insanity. Specifically, it puts in question the validity of any clear opposition between 'medical knowledge' and 'literary representations' of the insane at the turn of the nineteenth century. What does it mean to say that Mrs Fletcher's epitaph is 'literary', or, for that matter, that it is 'medical', when its ability to use a language shared by both popular fiction and medicine puts in question the demarcation of separate spheres of knowledge?

Marie Mulvey Roberts and Roy Porter have recently—and rightly—stressed the validity of approaching the history of eighteenth-century medicine and literature with 'a belief in an essential compatibility between the medical and the literary enterprises'.[48] By the start of Victoria's reign, however, that faith is no longer entirely

---

[47] Michel Foucault, *Histoire de la folie à l'âge classique* (Paris, 1972), trans. and abridged by Richard Howard, *Madness and Civilisation: A History of Insanity in the Age of Reason* (London, 1967); Jacques Derrida, 'Cogito et histoire de la folie', in Derrida, *L'Écriture et la différence* (Paris, 1967), trans. as 'Cogito and the History of Madness' by Alan Bass, in *Writing and Difference* (London, 1978), 31–63. Foucault responded to Derrida's reading in app. II to *Histoire de la folie*, new edition followed by 'Mon corps, ce papier, ce feu' (Paris, 1972), 583–603. On Foucault's influence on English history of psychiatry, see Arthur Still and Irving Velody (eds.), *Rewriting the History of Madness: Studies in Foucault's 'Histoire de la folie'* (London, 1992).

[48] Marie Mulvey Roberts and Roy Porter (eds.), *Literature and Medicine during the Eighteenth Century* (London, 1993), 1.

justified. Sarah Fletcher's epitaph provides a late expression of a culture in which literary and medical depictions of insanity shared much the same language and could still claim a comparable truth-value. To put the point more guardedly, the memorial speaks from a situation in which the nature of knowledge about mental and emotional processes, as about life and about death, is uncertain and in which no one authority or language has a prior claim to truth. During the next decades, however, medicine would make increasingly authoritative claims to an exclusive knowledge of the mind—claims backed by a more professional organization than ever before.

Numerous critics—most notably, in recent years, Foucault—have pointed to a fundamental conceptual change between the late eighteenth and early nineteenth centuries: a 'rearrangement of the hierarchy of knowledges' under which a juridical definition of selfhood and society gives way to a notion of the 'normal' or, by derivation, 'pathological' subject defined by the human sciences, and above all by clinical medicine.[49] It is easy to see how Sarah Fletcher's memorial might be interpreted in line with the Foucaultian model as marking an epistemic break in English thought, but to do so would almost certainly be to underestimate the variety of pressures shaping its appeal to a sentimental reading of female insanity. Written from outside the immediate purview of medicine, fiction, or, for that matter, the law, Sarah Fletcher's epitaph reflects the continuing compatibility and even interdependence of them all at the turn of the nineteenth century. No one discourse shapes this account of death, certainly not a juridical one, but also not in any simple sense a 'medical' or 'scientific' one. It draws on a language far more diffusive in its sources and its effects. Medicine has not yet become the primary arbitrator of knowledge about human life, but here intersects with, and is largely indistinguishable from, a language of sensibility and nervousness extensively cultivated by literate society. Indeed, the language of sentiment provides the means of rejecting knowledge (whether medical, legal, or religious) in favour of sympathy: a corrective

---

[49] Michel Foucault, *The Order of Things: An Archaeology of the Human Sciences*, trans. anon. (New York, 1970); Ernst Cassirer, *The Problem of Knowledge*, trans. William Woglom and Charles Hendel (New Haven, 1950), ch. 1; and, for discussion, Lawrence Rothfield, *Vital Signs: Medical Realism in Nineteenth-Century Fiction* (Princeton, 1992), 3–14.

force it would continue to exert in modified ways, despite the collapse of the late eighteenth-century culture of sensibility—as any reader of Dickens knows.

Even so, Sarah Fletcher's memorial could hardly have been written seventy years later. By 1865, when the Medico-Psychological Association took its name, medicine had laid its claim, albeit with profound internal and external conflicts, to a specialized knowledge of diseases of the mind. The increased profile of medical specialization in insanity was largely attributable to the growth of the county asylum system. The 1845 Lunatic Asylums Act finally obliged every county and borough in England and Wales to provide adequate institutional care at public expense for its lunatic paupers.[50] By the late 1860s the shortage of beds in county asylums required another Act of Parliament to license the construction of barrack-like asylums for the chronically mad. This unprecedented expansion in the asylum system underwrote the case for specialization in insanity. As Andrew Scull notes, not everyone within the medical establishment looked kindly on the mad-doctors, and many would have preferred to see them entirely separated from the rest of the medical profession.[51] Nevertheless, partly thanks to some aggressive self-promotion by men like John Conolly, George Mann Burrows, and Forbes Winslow, the specialization held its own.[52] 1865 may not have been the secure establishment of psychiatry (something achieved only at the very end of the century, if not even later) but it did mark the establishment of a visibly more organized, self-critical, and powerful professional body.

Janet Oppenheim has pointed out that the instability of nineteenth-century nomenclature for doctors specializing in mental pathology reflects ongoing uncertainty about the nature and prestige of specialization in diseases of the mind.[53] As fundamentally, however, the concerted *attention* given to nomenclature in the 1860s implies a growing social and professional self-awareness among this section of the medical community. As psychological medicine defines a role for itself, the development of its professional vocabulary is one of the clearest indices of the claim to an ever more

[50] Andrew Scull, *The Most Solitary of Afflictions: Madness and Society in Britain 1700–1900* (London, 1993), 155.
[51] Ibid. 250–1.
[52] Ibid. 251–66.
[53] Oppenheim, *'Shattered Nerves'*, 27.

exclusive authority. 'Psychiatry', 'psychiatric', 'psychiatrics', 'psychopathy', and 'psychopathology' all enter the language from German in the late 1840s; 'neuropathology', 'psychopathist', and 'psychiater' follow in the 1850s; the 1860s saw more additions to the language of mental science than any other period until the advent of psychoanalysis and sexology in the 1890s. Some terms were more lasting than others in a lengthy list which includes *psychopathologist* (1863), *psychosomatic* (1863), *psychometrist* (1864), *psycholatry* (1865), and *psychonomy* (1865).

As the 1857 *Encyclopaedia Britannica* noted in a greatly expanded entry on 'Mental Diseases', journals and asylum reports from leading asylums in England, France, Germany, and America had assisted the international exchange of practical and theoretical innovations in the diagnosis and treatment of insanity from the start of the century. Chief among the British improvements listed by the encyclopaedia were the extension of non-restraint and the establishment of the permanent Lunacy Commission after the reports of the 1815–16 Parliamentary Commission. By 1865 the medical profession could thus lay claim to its own language, its own professional body controlling standards of professional practice, its own publishing forums in which to pursue research into mental health and mental illness, and—not least—considerable public recognition of its successes. All these developments assisted a newly forceful distinction between professional and popular knowledge of insanity. In 1865 there was no longer a language which could be at once as effectively clinical and popular and as benignly emotive as the language of nervous sensibility had been. Lapidary inscriptions like the Fletcher epitaph, combining medical explanation with belletristic sentimentality, are rare after the turn of the nineteenth century, and when found their sentimentalism is usually self-consciously nostalgic. On the other hand, by 1865 even the most widely respected and sought-after doctors would still have no secure understanding of the human mind. The mental processes which could lead a deserted woman to suicide in 1799 would remain just as impenetrable to science seventy years on—after all, they remain substantially unsolved now. Despite the medical profession's advances, the shortfall between its hopes and its actual achievements, its claims to expertise and its ability to effect cures, was becoming ever more evident and a source of mounting public concern.

The repercussions for literature of psychological medicine's growing professionalization are complex. On the one hand, as the nineteenth century progressed, literature increasingly lacked the professional authority that medicine was claiming as its own, so that the novel's depictions of insanity could begin to seem outdated or too overtly dependent upon popular disseminations and dilutions of contemporary medical theory. On the other hand, medical claims to understand insanity were widely seen to be provisional, internally as well as externally disputed. Under such conditions, literature became the purveyor of 'lay' knowledge about insanity, yet, at the same time, preserved and even consolidated its right to speak about matters of emotional and psychological crisis for which medicine generally seemed to have provided no better answer.[54] In the hands of certain writers, the novel could become, in part, a forum from which medical claims to understand and treat insanity were openly challenged. The conventional narrative of a young woman driven insane by her misfortune in love proved a remarkably useful vehicle for such a challenge. Perennially popular, such stories possessed both the advantages and the drawbacks of familiarity. As an established convention, they had an immediate hold on people's recognition and credulity, but they also had that capacity of the 'merely' conventional to invite question.

The terms 'convention' and 'trope' have long been keywords in criticism of eighteenth- and early nineteenth-century literary and medical representations of female madness, recurring with particular frequency in feminist criticism. Certainly, there is much about insane women in writing of the period that is bound to strike present day readers as 'conventional' in the most pejorative sense of the word. Literary critics sympathetic to feminism have found the enduring popularity of fiction about women maddened by the desertion or death of their lovers peculiarly insidious. This narrative, more perhaps than any other, has supported the perception of femininity as a psychologically incomplete and characteristically unstable condition. Nevertheless, feminist criticism over the last

[54] There are other areas of medical practice whose claims to knowledge are equally debated in the 19th cent. and which possess a similarly complex relationship to literature. Mesmerism is one of the clearest examples. See Alison Winter, *The Island of Mesmeria: The Politics of Mesmerism in Early Victorian Britain*, doctoral dissertation, University of Cambridge, 1992.

two decades has also been founded on the conviction that madness carries with it an energy in excess of its conventional meanings: an energy which unsettles any merely conventional reading of the text in question.

In her 1975 article 'Women and Madness: The Critical Phallacy', Shoshana Felman analysed the representation of madness in a short story by Balzac in order to illustrate what she called the 'uncritical naïveté' of the 'realistic critic' who seeks in literature a single, univocal meaning conveyed in transparent language.[55] Balzac's 'Adieu' recounts the experience of a soldier who returns from the Napoleonic Wars to find that the woman he loved has gone mad in his absence and no longer recognizes him. All she can do is repeat the word she spoke to him in parting: 'Adieu'. The soldier's attempt to re-create the exact historical conditions of their last encounter in the hope of restoring her memory succeeds, but at the price of her life; she recognizes him, smiles, repeats 'Adieu' and dies. Felman's article noted that both the leading critical accounts of the short story in the early 1970s strikingly failed to mention either the woman or her madness, and treated the story instead as an exercise in historical realism—a faithful account of conditions during the Napoleonic Wars. The critics concerned had edited everything which might disrupt their own desire for unitary meaning, and failed to recognize that the story is itself a critique of the soldier's desire for such self-definition. Just as the soldier kills the madwoman in trying to force her to recognize him and make sense for him, the 'realistic critic' kills the play of meaning in the text by refusing to recognize the traces of violence, anguish, scandal, and insanity which it bears in the figure of the madwoman.

Sandra Gilbert and Susan Gubar's *The Madwoman in the Attic* (1979) is probably the best known of the flood of literary critical accounts of women and madness which followed in the late 1970s and 1980s.[56] Gilbert and Gubar were specifically concerned with

---

[55] Shoshana Felman, 'Women and Madness: The Critical Phallacy', *Diacritics*, 5/4 (1975), 2–10. Felman's article is in part a review of Phyllis Chesler's *Women and Madness* (New York, 1973). See also Jonathan Culler's discussion of Felman's essay in *On Deconstruction: Theory and Criticism after Structuralism* (Ithaca, NY, 1982; London, 1983), 62–3.

[56] Including Barbara Hill Rigney, *Madness and Sexual Politics in the Feminist Novel: Studies in Brontë, Woolf, Lessing and Atwood* (Madison, Wis., 1978); Shoshana Felman, *Writing and Madness (Literature/Philosophy/Psychoanalysis)*, trans. Martha Noel Evans with the author (Ithaca, NY, 1985); Patricia Vaeger,

the implications of madness as a theme in women's writing: if the madwoman disturbs the meaning of male-authored texts, how much more significant might such disruption be in works by women? Their chapter on 'The Woman Writer and the Anxiety of Authorship' provides a lucid summary of the terms on which they chose to celebrate the figure of the madwoman:

In projecting their anger and dis-ease into dreadful figures, creating dark doubles for themselves and their heroines, women writers are both identifying with and revising the self-definitions patriarchal culture has imposed on them. All the nineteenth- and twentieth-century literary women who evoke the female monster in their novels and poems alter her meaning by virtue of their own identification with her. For it is usually because she is in some sense imbued with interiority that the witch-monster-madwoman becomes so crucial an avatar of the writer's own self.[57]

The picture on the cover of the Yale University Press edition encapsulates Gilbert and Gubar's argument: in writing about madness, the woman writer draws the negative of herself, the dark, rebellious Other beneath the supposedly calm surface presented by writer and text. The hysteric writer produces a 'hysteric text', a term which came to enjoy wide currency in feminist criticism during the 1980s, and which was most thoroughly explored within psychoanalytic and post-structuralist frameworks by Hélène Cixous, Luce Irigaray, Mary Jacobus, Juliet Mitchell, and Alice Jardine.[58]

But how can the hysteric—a woman bereft of language—be 'an avatar of the writer's own self'? Luce Irigaray's feminist classic *This*

Honey-Mad Women: Emancipatory Strategies in Women's Writing (New York, 1988); Marilyn Valom, *Maternity, Mortality, and the Literature of Madness* (University Park, Penn., 1988); and, on Virginia Woolf, specifically, Stephen Trombley, '*All that Summer She was Mad': Virginia Woolf and her Doctors* (London, 1981); and Daniel Ferrer, *Virginia Woolf and the Madness of Language*, trans. Geoffrey Bennington and Rachel Bowlby (London, 1990).

[57] Sandra Gilbert and Susan Gubar, *The Madwoman in the Attic: The Woman Writer and the Nineteenth-Century Literary Imagination* (New Haven, 1979), 79.

[58] See Hélène Cixous and Catherine Clément, *The Newly Born Woman*, trans. Betsy Wing (Minneapolis, 1986), esp. 154, for Hélène Cixous's much cited claim that 'the hysteric is the typical woman in all her force'; Luce Irigaray, *This Sex which is not One*, trans. Catherine Porter (Ithaca, NY, 1985), 136–9; Juliet Mitchell, *Women: The Longest Revolution. Essays on Feminism, Literature and Psychoanalysis* (London, 1984), 287–94; Alice Jardine, *Gynesis: Configurations of Woman and Modernity* (Ithaca, NY, 1985), 178–207; and Mary Jacobus, *Reading Woman: Essays in Feminist Criticism* (London, 1986), 197–274. See also, Elaine Showalter, 'On Hysterical Narrative', *Narrative*, 1 (1993), 24–35.

Sex which is not One (1977) specifically defended the ability of hysteria to 'speak' through its own silence:

it speaks in the mode of a paralyzed gestural faculty, of an impossible and also a forbidden speech . . . it speaks as *symptoms* of an 'it can't speak to or about itself' . . . Both mutism and mimicry are then left to hysteria. Hysteria is silent and at the same time it mimes. And—how could it be otherwise—miming/reproducing a language that is not its own, masculine language, it caricatures and deforms that language: it 'lies,' it 'deceives,' as women have always been reputed to do.[59]

This is hysteria as the breaking-point of language, the mute and painful dramatization of an exclusion from speech. The same conviction that hysteria embodies an otherwise inarticulable distress underlay Juliet Mitchell's declaration in 1984 that 'the woman novelist must be an hysteric. Hysteria is the woman's simultaneous acceptance and refusal of the organisation of sexuality under patriarchal capitalism. It is simultaneously what a woman can do both to be feminine and to refuse femininity, within patriarchal discourse.'[60] Like many branches of literary theory, feminist criticism during the 1970s and 1980s seemed to have adopted a vocabulary of extremes, and one in real danger of devaluation. Whether it was the hysteric text or (from Deleuze and Guattari, and Irigaray) 'schizophrenic writing',[61] the appropriation of the terms of insanity marked a challenge to contemporary psychiatry and psychoanalysis, and to traditional literary critical methodologies. But is it justifiable to use madness as a symbolic gesture towards everything the text or the writer will not or cannot say? Doesn't this desire to give the hysteric a voice become as coercive of madness as Foucault tells us post-Enlightenment philosophy has always been?

The attraction of the madwoman as a deconstructive key to all texts produced under patriarchy is evident: as a figure of rage, without power to alleviate her suffering or to express it in terms which make sense to society, she sums up virtually everything feminism might wish to say about the suppression of women's

---

[59] Irigaray, *This Sex which is not One*, 136–7.

[60] Mitchell, *Women: The Longest Revolution*, 289–90.

[61] See Gilles Deleuze and Félix Guattari, *L'Anti-Oedipe: Capitalisme et schizophrénie* (Paris, 1972), *Milles Plateaux* (Paris, 1980), and *Semiotext(e)* (New York, 1983); and Luce Irigaray, *Parler n'est jamais neutre* (Paris, 1985).

speech. But if Gilbert and Gubar demonstrated just how dramatic the effects of liberating the madwoman from the attic could be for literary criticism, they also laid bare the dangers of romanticizing sickness and alienation. Mary Jacobus, reviewing *The Madwoman in the Attic*, recognized the engaging energy of its rereadings of the canon but refused to endorse a critical strategy which she saw as, in the end, reductive. In its eagerness to liberate the covert or unacknowledged aggression of the woman writer, she argued, *The Madwoman in the Attic* becomes 'not just revisionary but diagnostic'. All women's writing is made to reveal the same myth of origins. Jane Austen, Emily Dickinson, Charlotte Brontë, George Eliot become alike, because they all testify in the same way to the psychic stress induced in any woman who takes up her pen in a patriarchal society.[62]

Mary Jacobus's objections are representative of a deep-seated unease still running through feminist literary theory over the nature and uses of the 'woman and madness' conjunction. The charges laid against feminist appropriations of the hysteric as the quintessential woman are several. Most obviously, the celebration of hysteria risks culpably reinscribing women in precisely the debilitating gender constructions feminism might hope to free them from, equating them once again with irrationality, silence, nature, and body. It also ignores the medical history of women's relationship to madness—hysteria and depression have been far from liberating for most of the women who, from the eighteenth century, have made up the majority of patients in mental asylums.[63] At times it seems that we have come to *expect* madness of our women writers, as if in the belief that nothing less than madness can testify to a proper sensibility of female oppression. Catherine Gallagher has remarked that writers like Aphra Behn who take patriarchy 'too much in [their] stride' and who speak more of resilience than of depression or 'psychic damage' have sometimes appeared to perplex and embarrass feminist criticism.[64]

---

[62] Mary Jacobus, Review of *The Madwoman in the Attic* and *Shakespeare's Sisters*, *Signs*, 6 (1981), 517–23: 520; cf. Catherine Clément's response to Hélène Cixous, *The Newly Born Woman*, 156–7.

[63] See Showalter, *The Female Malady*, 3, 52.

[64] Catherine Gallagher, 'The Networking Muse: Aphra Behn as Heroine of Frankness and Self-Discovery', *Times Literary Supplement*, 10 Sept. 1993, 3–4: 3.

Right at the start of the 1970s debate, Shoshana Felman noted the complementarity and yet the fundamental incompatibility of the methodologies pursued in two feminist studies of madness: Phyllis Chesler's *Women and Madness* and Luce Irigaray's *Speculum*. The difference between Chesler's materialist endeavour to let women's writing about insanity 'speak for itself' and Irigaray's theoretical critique of the exclusion of women and madness from language and from philosophy neatly reflects a division which would at one time have been described as Anglo-American versus French feminism. Though the map has changed, the underlying tensions have not been resolved. Felman's answer to the suppression of the feminine she discovered in reading Balzac's critics was 'nothing less than to "re-invent" language, to *re-learn how to speak*'—and the same answer has kept coming back in feminist writing about madness (notably in Drucilla Cornell's 1991 *Beyond Accommodation*). But it is here that the term 'convention' needs to re-enter the discussion. Neither the view of literature as a place where we can hear women speaking for themselves, nor the understanding of literature as a site where meaning—including the meaning of femininity—is constantly being reinvented, pays sufficient attention to the nature of literary convention and the degree to which it can be resisted, reformed, or, for that matter, rejected.

Mary Wollstonecraft has long been a key figure within Anglo-American debates about the nature and the limits of convention. Like Sarah Fletcher, she died at the very end of the eighteenth century; like Sarah Fletcher, again, she was driven to attempt suicide. Unlike Sarah Fletcher, she wrote about women and madness, and her writing reveals what it was possible to say and think at a time when sentimentalism was still the dominant influence on novelistic representations of female insanity, and when criticism of medical attitudes to women was barely making itself heard as yet. All Wollstonecraft's novels reflect upon the mental health of women in her society. In her first, *Mary* (1788), she explored at length the psychological freedoms as well as the dangers which the culture of sensibility presented for women. *The Wrongs of Woman; or, Maria*, which remained unfinished at the time of the author's death in 1797, is a much bleaker account of the painful reduction of women's emotional and intellectual life within that same culture. It was published the following year, tactfully edited by her husband, William Godwin, who left it in fragmentary form with only slight additions to establish the flow of the narrative.[65]

Maria opens with its heroine incarcerated in a madhouse by her profligate and debt-burdened husband, who has made an aggressive bid for solvency by seizing her inheritance from an uncle. Wollstonecraft makes immediate and crucial changes in the way Mackenzie, Cowper, and their imitators in the periodicals represented the mental distress of women disappointed in love. The novel is overtly polemical, prefaced by the author's declaration that she aims to exhibit 'the misery and oppression, peculiar to women, that arise out of the partial laws and customs of society'.[66] 'Matrimonial despotism of heart and conduct' are to her 'the peculiar Wrongs of Woman, because they degrade the mind'.[67] Rejecting the assumption that insanity follows naturally for a woman deceived by her lover, and rejecting, more generally, the sentimental valuation of madness as a quintessentially feminine condition, her final novel subjects the intellectual and emotional disempowerment of women to a searing economic and social critique.[68]

Wollstonecraft's crucial break with sentimental convention is to make Maria's supposedly deranged consciousness the novel's focus. Although the name immediately evokes Laurence Sterne's madwoman, this Maria is no picturesque object for the reader's gratified contemplation but an angry and articulate woman. Maria puts a forceful case against sentimentalism, insisting that her culture's predilection for stories and images of overwrought feminine sensibility has a material effect on women's lives, discouraging them from the healthy development of the intellectual faculties and laying them open to exploitation by men. Wollstonecraft's heroine may have been deceived and abandoned by her lover, but her 'madness' is entirely fabricated by him. Unlike Sterne's tender maid, who wanders the countryside living off the charity of her

[65] Mary Wollstonecraft, *The Wrongs of Woman; or, Maria. A Fragment* (1798), in Janet Todd and Marilyn Butler (eds.), *The Works of Mary Wollstonecraft*, 7 vols. (London, 1989), i.

[66] *The Works of Mary Wollstonecraft*, i. 83.

[67] Ibid. i. 84.

[68] Several feminist critics have argued that the critique of romanticism deprives it of a sense of, or, more accurately, a language for the positive expression of female sexuality. For example, Cora Kaplan, *Sea Changes: Culture and Feminism* (London, 1986), 34–50. Janet Todd defends Wollstonecraft, on the grounds that 'Women are dangerous' in Wollstonecraft's analysis, 'mainly as the objects of desire'; *Feminist Literary History: A Defence* (Oxford, 1988), 108.

neighbours, this Maria is placed in the public domain of madness, the city asylum, where 'wretched minds' are managed with an eye to commercial profit.[69] Resisting the sentimentalists' too easy yoking of sympathy and charity, *The Wrongs of Woman* recognizes and fights the trade in women's minds and bodies. Prostitution is a recurrent theme and becomes, like the asylum itself, a charged symbol of economic and sexual exploitation.

Feminist readings of *The Wrongs of Woman* have, however, deeply disagreed about the extent to which Wollstonecraft succeeds in freeing herself from the limitations of sentimental convention. Margaret Anne Doody and Mary Jacobus have both staunchly defended Wollstonecraft's radicalism. In their view, *The Wrongs of Woman* uses the representation and the language of maddened sensibility to protest against the values of Enlightenment reason. Other critics have disputed that claim. Janet Todd, for example, in *Feminist Literary History: A Defence* (1989) writes:

It seems to me that in this reading theory has overwhelmed history and genre. Penetrating the surface of the text, Jacobus has ignored the surface of history, of class and politics, for example, to find critical excess and turn her subject into a kind of shadowy Virginia Woolf. Stimulating though the description of madness is—the picture of Wollstonecraft contemplating insanity, like an early Jane Eyre after Gilbert and Gubar—it should not be isolated from the specific historical context that makes of madness a stance of the period, a trope of sentimental fiction, as well as an individual posture.[70]

In Janet Todd's view, the use of madness as a sign that the text protests against its ideological context involved emptying madness of representation, making it 'pure' metaphor: a politically innocent symbol of division and alienation. In pushing for a recognition that madness might function as 'a trope of sentimental fiction, as well as an individual posture' she was asking for a recognition that madness is more than a black hole in a text; that it is structured in language and carries with it implied or overt codes of interpretation. The question then becomes one of whether a particular text merely reproduces popular images of madness or whether it addresses its own status as convention.

---

[69] Philip Martin has drawn attention to this distinction; *Mad Women in Romantic Writing*, 21.
[70] Todd, *Feminist Literary History*, 106.

Mary Wollstonecraft provides an apt starting-point for a discussion of love-madness in the nineteenth century because she so clearly confronts the manipulability of the madwoman as a sentimental icon. In freeing Maria from the terms which hold a sentimental reading in place, Wollstonecraft makes her a focal symbol for wider political and social concerns, anticipating the debate within twentieth-century feminism by nearly 200 years. At the same time, Wollstonecraft repeatedly acknowledges the strain of making her resistance to convention felt. Her novel derives both anxiety and added force from the knowledge that its very transgressiveness against the codes of sensibility may allow it to be readily reassimilated within the very model it seeks to unsettle. 'The Wrongs of Woman, like the wrongs of the oppressed part of mankind, may be deemed necessary by their oppressors: but surely there are a few, who will dare to advance before the improvement of the age, and grant that my sketches are not the abortion of a distempered fancy, or the strong delineations of a wounded heart.'[71] With that appeal, the novel seems to be seeking to pre-empt the criticism most likely to be raised against it by a readership schooled to recognize women's anger and rebellion as symptomatic of emotional and sexual derangement. The sentimental movement produced, as Wollstonecraft knew, not only a series of writing conventions but a habit of reading which expected those conventions to be fulfilled. *The Wrongs of Woman* is written in the knowledge that such habitually narrowed reading practices have the power to defuse any writer's attempt to rethink or redirect the figure of the madwoman. The attraction of the conventions surrounding the madwoman, she seems to suggest, may be such that they will always forestall her appropriation in the service of a radical social and literary critique. The attitude of the judge who finally dismisses Maria's divorce claim indicates Wollstonecraft's fears about the way her novel would be read: 'For his part, he had always determined to oppose all innovation, and the new-fangled notions which incroached on the good old rules of conduct.'[72]

Wollstonecraft is typically depicted as a woman whose intense vision of a sexually equitable society was far in advance of her time. But, while she was undeniably an outstanding figure in feminist

---

[71] *The Works of Mary Wollstonecraft*, i. 83.
[72] Ibid. i. 181.

literary history, she was far from alone in wanting to rethink contemporary representations of female insanity. As *Love's Madness* aims to show, stories about women driven mad by the death or infidelity of their lovers were so thoroughly conventional that they obliged writers to confront their literariness and to make particular efforts at originality. In that respect, the love-mad woman brings not an excess of meaning but a reduction of meaning. Almost all the writers who pick up this convention in the nineteenth century do so with a degree of distance, a felt need to say something *more* with it or through it. Love-mad women inspired a great deal of tawdry writing, but they also spurred many novelists into remarkable innovation. For medical writers in search of a more effective mode of treating the insane, they were beginning to seem rather less tractable.

2

Love-Mad Women and the Rhetoric of Gentlemanly Medicine

WHEN Alexander Morison addressed the subject of love and madness in his *Outlines of Lectures on Mental Diseases* (1825), he did so in terms cautiously evocative of the late eighteenth-century cult of sensibility. 'Love', he told his students, 'produces febrile symptoms, and increased sensibility. when hopeless—sometimes insanity.'[1] Morison seems to have extemporized fairly freely from his notes when he lectured, and in this case the reference is so brief that it is unclear whether he distinguished between men and women at all in his discussion. However, the slightness of Morison's allusion here to the most affecting literary type of insanity was compensated for in later editions. The greatly expanded 1848 *Outlines of Lectures on the Nature, Causes, and Treatment of Insanity* not only clarified his views on female insanity in general, but gave considerably more space to 'disappointed love'. A table informs the reader that out of 562 cases of insanity treated at Bethlehem Hospital (the date and period concerned are not stated), twenty-five were the result of disappointed love. Of these, twenty were female, five were male.[2] The text remains reticent, however, about the conclusions to be drawn from this free-floating statistic. Morison begins by referring to the popular perception that love-madness was a disease of young women:

According to Zimmerman, the passion of love makes girls mad; jealousy, women mad; and pride, men mad. The former passion, that of love, has been a fruitful source of insanity in all ages, and jealousy and ambition not less so.[3]

[1] Alexander Morison, *Outlines of Lectures on Mental Diseases* (Edinburgh, 1825), 62.
[2] Sir Alexander Morison, *Outlines of Lectures on the Nature, Causes and Treatment of Insanity*, 4th edn., ed. Thomas Coutts Morison (London, 1848), 319.
[3] Ibid. 314.

The second sentence is notably less pat than the first, and, having quoted received opinion, Morison promptly contradicts it by introducing an example of male insanity from love:

The emotion of joy has been sometimes considered as a cause of insanity, but this, I think, arises from not sufficiently inquiring into all the particulars of the case. For instance, a young man through the interest of a friend, obtained a very desirable appointment; but to profit by it, his immediate departure was necessary. Shortly afterwards he became insane, and it was generally believed that joy had turned his head; on more particular inquiry, however, it appeared that grief, on being obliged to part from a young lady, to whom he was ardently attached, was the real cause.[4]

These were not Morison's last words on love-madness, but his desire not to perpetuate sentimental myths is evident. What, in fact, was the relationship between fictional and medical depictions of insanity in early nineteenth-century England? How and to what end did doctors generally allude to literary models of psychology in their writing, and how particularly did they respond to the sentimental cult of love-mad women?

The similarities between representations of female madness in fiction and in medicine during the eighteenth and nineteenth centuries have proved a fruitful area for interdisciplinary research. Both Sander Gilman and Elaine Showalter have drawn attention to the asylum photography of Hugh W. Diamond, identifying striking iconographic echoes of literature's madwomen in these supposedly 'scientific' images: a sentimentalized Ophelia dressed in white, with flowers in her hair; a group of deluded queens reminiscent of Lewis Carroll's dotty White Queen in *Through the Looking-Glass*.[5] More generally, there is now an extensive body of critical literature concerned with the parallels between the figuring of madwomen in popular fiction and the representation of female patients in medical texts and asylum records.[6]

---

[4] Morison, *Outlines of Lectures on the Nature ... of Insanity* (1848), 314–15.

[5] Elaine Showalter, *The Female Malady: Women, Madness, and English Culture 1830–1980* (London, 1987), 86–97; Sander L. Gilman (ed.), *The Faces of Madness: Hugh W. Diamond and the Origin of Psychiatric Photography* (New York, 1976); Gilman, *Seeing the Insane: A Cultural History of Madness and Art in the Western World* (New York, 1982), 127, cf. 208.

[6] Notably Jenny Bourne Taylor, *In the Secret Theatre of Home: Wilkie Collins, Sensation Narrative, and Nineteenth-Century Psychology* (London, 1988); and Sally Shuttleworth, ' "The Surveillance of a Sleepless Eye": The Constitution of Neurosis in *Villette*', in George Levine (ed.), *One Culture: Essays in Science and Literature*

For literary critics, medical history has provided a useful means of grounding fiction in experience, enabling literature's hysterics to be brought into sisterhood with the inhabitants of the real asylums. Alongside Scott's Lucy Ashton and Charlotte Brontë's Bertha Mason, research into the history of medicine has permitted literary criticism to place the actual subjects of early nineteenth-century psychological treatment: the case histories from asylum records, the faces Diamond photographed. Medical literature of the eighteenth and nineteenth centuries has been repeatedly scrutinized for its pronouncements on mania, on *hysteria nervosa*, on puerperal insanity, examined (and, for the most part, denounced) in terms of the ideological assumptions it shares with its contemporary fiction.[7] The medical diagnosis and the literary depiction of madness have emerged as substantially the same: difference of discourse conceals a shared political unconscious.

Any difficulty experienced in making the transition between medicine and popular fiction has usually been smoothed over as rapidly as possible. Lillian Feder in *Madness in Literature* (1980), for example, notes the problems of interpretation posed by historical data, yet finds no difficulty in seeing literature as an 'echo' of contemporary philosophical and medical theories on mental disorder.[8] For Philip Martin, in *Mad Women in Romantic Writing* (1987), the difference is in intensity of scrutiny rather than in kind: 'While women's madness and derangement were being written into the literary texts of the Romantic period and beyond, it was receiving more detailed and direct treatment in the medical and psychiatric writings of the same period.'[9] Elaine Showalter's *The Female Malady* puts the case for the 'one culture' model most forcefully of all: in her account medical theory and medical practice endorsed and perpetuated the gendering of madness as a quintessentially feminine condition, and, as the nineteenth century progressed, gave

(Madison, Wis., 1987). See also John Mullan, *Sentiment and Sociability: The Language of Feeling in the Eighteenth Century* (Oxford, 1988), and Allan Ingram, *The Madhouse of Language: Writing and Reading Madness in the Eighteenth Century* (London, 1991).

[7] See e.g. Sally Shuttleworth, 'Ideologies of Bourgeois Motherhood in the Mid-Victorian Era', in Linda Shires (ed.), *Rewriting the Victorians: Theory, History, and the Politics of Gender* (New York, 1992).

[8] Lillian Feder, *Madness in Literature* (Princeton, 1980), 29, 151. Feder is referring to 17th- and 18th-cent. literature in this instance.

[9] Philip Martin, *Mad Women in Romantic Writing* (Brighton, 1987), 28.

'the full weight of scientific confirmation to narrow Victorian ideals of femininity'.[10] The phrase gives expression to a widespread tendency to see medicine as simply 'confirming' narrow ideals already in circulation in British culture. But while it is undeniably the case that the history of women's madness in Britain (as on the Continent and in America) contains ample evidence of brutal misogyny, not enough questions are being asked about the nature of the relations between medicine and literature. When Showalter writes of Crazy Jane as 'a fascinating demonstration of the traffic between cultural images and psychiatric ideologies',[11] she suggests that, if 'psychiatry' and 'culture' are separate realms, images and ideologies nevertheless flow smoothly between them.

Neither, surprisingly, has there been much pressure from historians of medicine to question such easy cross-reference between literary and medical writing. Both literary critics and medical historians have tended to assume that the evidence of nineteenth-century medical and fictional texts is bound to be complementary. W. F. Bynum and Michael Neve's 1985 essay 'Hamlet on the Couch', which remains one of the most detailed studies of the way in which medicine has appropriated literary representations of insanity, conducts a thorough survey of nineteenth- and twentieth-century medical commentaries on *Hamlet*, in order to chart changing psychiatric and psychoanalytic definitions of sanity and madness. Their list of nineteenth-century medical writers who address the question of Hamlet's insanity is extensive: John Conolly, John Bucknill, Henry Maudsley, Forbes Winslow, in England; Isaac Ray, Amariah Brigham, A. O. Kellog, Ludwig Jekels, in America; and, on the Continent, Cesare Lombroso, A. Delbrück, Heinrich Laehr, and H. Turek. *Why* Shakespeare should have been taken up by these writers as an appropriate subject for medical discussion remains obscure; and here again the general emphasis in interdisciplinary studies is on complementarity.[12]

While literary critics have turned to medical history to ground their readings of fiction's mad people in the lived experience of

[10] Showalter, *The Female Malady*, 121–2, and 'Representing Ophelia: Women, Madness, and the Responsibilities of Feminist Criticism', in Patricia Parker and Geoffrey Hartman (eds.), *Shakespeare and the Question of Theory* (London, 1985).
[11] Ibid. 6–7.
[12] In W. F. Bynum *et al.* (eds.), *The Anatomy of Madness: Essays in the History of Psychiatry*, 3 vols. (London, 1985–8), i. 289–304.

eighteenth- and nineteenth-century men and women, medical historians have responded in kind. Autobiographical writings by mad
writers are commonly invoked (especially Swift, Blake, Cowper,
Clare), but historians have also looked more widely to literature as
one of the most productive sources for a social history of madness
to complement the institutional history of medical science.[13] Indeed,
autobiography, biography, and literature tend to merge as sources
for social histories of madness, so that questions of authorship,
experience, and genre seem irrelevant even where they may be
crucial. Warned by Foucault's insistence that the experience of
madness has, since the Enlightenment, been denied expression in a
language shaped by the demands of reason, medical historians have
been anxious to find a means of moving beyond the official institutional records and medical textbooks to retrieve something of
what it felt like to be mad in the past. A strictly Foucaultian
approach would insist that the literary and pictorial self-expres-
sions of the insane are just as much shaped by institutional, cultural, and linguistic pressures as are the writings of doctors.
Nevertheless, for those historians wanting to take account of more
than just the official records of the insane, literature and art have
come to seem the only legitimate means of approaching 'madness
itself', in Foucault's much disputed phrase.[14]

Social histories of madness reveal a set of tensions which underlie, less visibly, the writing of any social history of medicine. Pa-
tient-centred histories of physical illness are driven by ready
sympathy (we are all sick at some point in our lives), but patient-
centred histories of madness offer a less attractive identification.
When Roy Porter begins his *Mind Forg'd Manacles* (1987) with a
series of literary and autobiographical quotations prompting a
recognition of our own possible slippage into madness—Polonius

[13] See e.g. Roy Porter, *Mind Forg'd Manacles: A History of Madness in England
from the Restoration to the Regency* (London, 1987), 92–107 and *passim*; Porter, *A
Social History of Madness: Stories of the Insane* (London, 1987); Andrew Scull (ed.),
*Madhouses, Mad-Doctors, and Madmen: The Social History of Psychiatry in the
Victorian Era* (London, 1981). Also William Llewellyn Parry-Jones, *The Trade in
Lunacy: A Study of Private Madhouses in England in the Eighteenth and Nineteenth
Centuries* (London, 1971); Max Byrd, *Visits to Bedlam* (Columbia, SC, 1974); and
Michael V. DePorte, *Nightmares and Hobbyhorses: Swift, Sterne, and Augustan
Ideas of Madness* (San Marino, Calif., 1974) amongst others.
[14] See Nikolas Rose, 'Of Madness Itself: *Histoire de la folie* and the Object of
Psychiatric History', in Arthur Still and Irving Velody (eds.), *Rewriting the History
of Madness: Studies in Foucault's 'Histoire de la folie'* (London, 1992).

proposing that to define true madness is to be mad; young Charles Darwin declaring 'My father says . . . that everybody is insane at some time'[15]—we might well ask how threatened we are really supposed to feel. *Mind Forg'd Manacles* is ultimately a reassuring book, in that it renders the mad so bizarre and, in many instances, so comic that it comforts us with the covert implication that, although the line between the sane and the insane may be slippery, although it may have been drawn differently and drawn wrongly in the past, it is nevertheless there. To turn to literature in search of a historical procedure which can recover 'The Voice of the Mad' is as odd as literary critics' appeal to the history of medicine as an autonomous body of facts. Instead of exploring the complex ways in which medical theory and practice remained independent of, yet also partly shaped by, literary models of insanity, historians have used literature too simply as a social complement to the institutional story.

The questions provoked by recent interdisciplinary accounts of the representation of female insanity in nineteenth-century England have much broader implications for the history of madness and for literary criticism. Were the links between nineteenth-century medicine and literature necessarily so direct and untroubled? What is implied when we claim to have located traffic in images between the wider literary culture and medical science? Did medical writing simply reproduce and foster popular images and narratives of madness, albeit within an increasingly professionalized and élitist vocabulary? And was that writing always so coercively obedient to narrow Georgian and Victorian ideals in its politics, especially its gender politics, or did doctors also have other stories to tell?

These questions can be pursued through a selection of works which have come to be regarded as central to the development of nineteenth-century psychological medicine: Joseph Mason Cox's *Practical Observations on Insanity* (1804), John Conolly's *An Inquiry Concerning the Indications of Insanity* (1830), Henry Maudsley's *The Physiology and Pathology of the Mind* (1867), and, almost spanning them, the several editions of Alexander Morison's *Lectures on Insanity*. To turn again to Cox, Conolly, Maudsley, and Morison is to run the risk of reinforcing a 'psychiatric' canon whose revision has only recently begun in

---

[15] Porter, *Mind Forg'd Manacles*, 1.

earnest.[16] Rereading the relationship between the 'literary' and the 'scientific' with primary reference to a small selection of physicians cannot do justice to the more complex picture presented by domains of medical practice which have traditionally remained tangential to the history of psychiatry or been quietly subsumed within it: phrenology, craniology, mesmerism, and spiritualism, for example.[17] But, precisely because these men have helped to entrench the notion of a 'psychiatric science' whose relations with a 'wider culture' can be meaningfully talked about, it is necessary at this juncture to reread their writings and to question the way in which they thought about their relationship to literature.

Cox, Conolly, Maudsley, and Morison have had considerable influence on the way in which other disciplines—in this case, specifically literary criticism—have thought about relations between medical science and wider English culture in the period. Conolly and Maudsley stand respectively for the popularization of non-restraint and the emergence of 'Darwinian psychiatry'. Both are frequently cited as men whose medical theories were aided and prejudiced by literary stereotypes. Similarly, Morison's highly influential work on the physiognomy of the insane has been seen as strong evidence of Victorian medicine's attachment to literary models of female insanity and to a coercively normative concept of femininity. Cox's writing is less centrally important to the history of medical thought, but his vigorous advocacy of physical treatments for the insane was widely acclaimed in its day and his performative and engaging treatise has lent itself particularly well to a demonstration of images and narratives travelling with apparent ease between popular fiction and medicine. Yet, on closer reading, far from simply confirming narrow nineteenth-century ideals of femininity, each of these works exhibits considerable ambivalence towards the most pervasive stereotype of female insanity, the love-mad woman; and rather than indicating unproblematic continuities between

[16] A more detailed discussion of the psychiatric canon and its implications can be found in Helen Small, ' "In the Guise of Science": Literature and the Rhetoric of Nineteenth-Century English Psychiatry', *History of the Human Sciences*, 7 (1994), 31–3.

[17] e.g. Clara Gallini, *La sonnambula meravigliosa: Magnetismo e ipnotismo nell'Ottocento italiano* (Milan, 1983); Barbara Spackman, *Decadent Genealogies: The Rhetoric of Sickness from Baudelaire to D'Annunzio* (Ithaca, NY, 1989); Alex Owen, *The Darkened Room: Women, Power, and Spiritualism in Late Nineteenth-Century England* (London, 1989).

medical and literary culture, their use of literary allusion more generally is shaped by highly specific professional pressures.

Joseph Mason Cox and Literary Madwomen

The late Victorian psychologist D. H. Tuke called Joseph Mason Cox's *Practical Observations* 'the best medical treatise of the day on insanity'.[18] Cox is identified in Hunter and Macalpine's *Three Hundred Years of Psychiatry* as 'the first regularly qualified physician and author of a treatise on insanity who studied medicine in order to specialise in mental diseases'.[19] His medical education took him to Edinburgh, Paris, and Leiden, and he returned to England in 1788 to take over the management of his father's asylum at Fishponds, near Bristol. The *Practical Observations* were written after sixteen years' experience of private practice there. They are, as the title suggests, a series of case histories and illustrations intended to clarify for the benefit of medical students the nature and causes of insanity; and the book proved immediately popular in a growing market for books about the treatment of the mad. It quickly went through three editions in England and one in America, and was translated into both French and German.[20]

Practical Observations begins by delivering a round rebuke to the 'majority of medical authors', who 'have seemed more anxious to display their ingenuity and the result of their abstruse speculations than to furnish the inquiring student with a plain practical manual' (p. vi).[21] *Practical Observations* aims, as the title suggests, to redirect the doctor from abstract to human subjects, advocating not a theorizing inward gaze but a more 'hands on' method—and Cox's medical practice is nothing if not hands on. Despite its

---

[18] Quoted in Richard Hunter and Ida Macalpine, *Three Hundred Years of Psychiatry, 1535–1860* (London, 1963), 594.

[19] Ibid. 594.

[20] Andrew Scull, *The Most Solitary of Afflictions: Madness and Society in Britain 1700–1900* (London, 1993), 74.

[21] See also the advertisement for Cox's book, repr. p. iii of the book. Cox's rebuke fits a pattern for treatises on madness from the late 18th cent. right through the 19th. For similar examples, see Thomas Trotter, *A View of the Nervous Temperament* (London, 1807), p. xvi; J. C. Prichard, *A Treatise on Insanity* (offprinted from *The Cyclopaedia of Practical Medicine*) (London, 1833), 3–4; W. A. F. Browne, *What Asylums Were, Are, and Ought to Be* (Edinburgh, 1837), 3; and, for a late example, J. H. Tuke, *The Insanity of Over-Exertion of the Brain* (Edinburgh, 1894), 1–3.

emphasis on moral as well as medical management, the book is noteworthy for its popularization of the 'swinging' treatment, which it describes extensively, invoking numerous case-studies to demonstrate its effectiveness. Beyond this, the treatise promotes the standard formidable range of methods for dealing with recalcitrant cases of insanity: purgatives, emetics, blisters on the shaven scalp, restraint (for violent cases), as well as the less common practice of inoculating a patient with smallpox. Cox asserts that two diseases cannot inhabit the body at the same time ('no fact in medicine is more completely established'; 2) and claims that smallpox may therefore be used to drive out insanity.

Yet, for all its claims to be plain and practical, this manual is substantially devoted to theoretical speculation; or, more accurately, it struggles to negotiate between medical practice and a theory of insanity which must remain substantially speculative.[22] Its first thirty pages consist of an attempt to account for the origins of insanity, describe its characteristics, and determine the best means of curing it. They produce—and this point is endemic to treatises on madness in the period—a proliferation of causes on the one hand, and symptoms on the other, under whose pressure the goal of definition constantly recedes. The problem is complicated by the tendency of both causes and symptoms to oscillate between types of insanity which Cox would like to think of as distinct (violence can denote both mania and depression; venality can issue in either melancholia or depression). References to other medical sources offer little help, for they feature in most instances only to be dismissed as inadequate to the problem in hand.[23] Contemplating a disease whose causes range from venereal and alcoholic indulgence to excess of religious fervour, and with a suffering body whose symptoms range from violence to torpidity, the treatise constantly finds itself drawn into the kind of abstruse theorization which it is

[22] On speculation and insanity, see Small, 'In the Guise of Science', 34.

[23] e.g. *Practical Observations*, 18, regarding Cullen's definition of madness as 'Delirium sine febre'. Significantly, the only sources Cox footnotes with agreement are in Latin. See ibid., p. vi on Lorry's *De Melancholia* (1765), or p. 2 on Mead's *Monita* (1751). The practice of quoting in Latin was rapidly fading out in medical writing during the early 19th cent. Cox seems to have recourse to it at the points where his assertions are particularly tenuous. Lorry is used to support a claim that the medical establishment has lost what 'our ancestors once knew'; Mead is used to support the claim that insanity so takes over the body as to secure it from all other diseases, and hence to support the practice of inoculation with smallpox in the hope of driving out insanity (the logic is clearly faulty, to say nothing of the practice).

at pains to avoid. Thus, it is not surprising to find the physician more than once abandoning a train of inquiry that threatens to lead him into the madness of speculation. A series of florid gestures register the impenetrability of the subject: speaking of 'connate predisposition', Cox concedes 'It is hidden among the arcana of Nature, beyond the reach of human comprehension, on what these hereditary peculiarities depend' (5); and of the 'Proximate Cause' of insanity, 'I forbear to enter upon a topic, on which, like my predecessors, I could offer nothing but vague conjecture and speculation' (14).

Cox has one powerful resource in controlling a text which threatens to follow its subject into madness. The case-study provides him with a narrative logic which the general discussion of insanity denied him: the history of the disease, its symptoms, the mode of treatment, the successful cure. The individual experience permits him to find as ontogeny a history which could not be written as phylogeny; and grouped together, these self-contained narratives work to authenticate and control a subject which until now has been unmanageable. Cox uses the case-study format in a highly directed manner, with the doctor's control on constant display, as the first group of twelve case-studies reveals. It is introduced to illustrate a mode of treatment which Cox describes as

pious frauds: as when one simple erroneous idea stamps the character of the disease, depriving the affected party of the common enjoyments of society, though capable of reasoning with propriety, perhaps with ingenuity, on every subject not connected with that of his hallucination, the correction of which has resisted our very best exertions, and where there is no obvious corporeal indisposition, it certainly is allowable to try the effect of certain deceptions, contrived to make strong impressions on the senses, by means of unexpected, unusual, striking, or apparently supernatural agents; such as after waking the party from sleep, either suddenly or by a gradual process, by imitated thunder or soft music, according to the peculiarity of the case, combating the erroneous deranged notion, either by some pointed sentence, or signs executed in phosphorus upon the wall of the bed chamber,[24] or by some tale, assertion, or reasoning; by one in the

[24] Cox's suggested use of 'signs executed in phosphorus upon the wall of the bed chamber' is picked up, and perhaps satirized, in Charles Maturin's *Melmoth the Wanderer* (1820), where the inmates of a Spanish monastery attempt to subordinate a reluctant member of the order, or to drive him mad, by writing warnings of damnation in phosphorus on his cell walls. See Maturin, *Melmoth the Wanderer: A Tale*, ed. Douglas Grant (London, 1968), 153.

character of an angel, prophet, or devil: but the actor in this drama must possess much skill, and be very perfect in his part. (28–9)

The sheer length of the sentence, heaping qualification upon quali-fication, is perhaps indicative of some anxiety about the reception of these methods. Certainly Cox has reason to be anxious, and though he encourages other members of the medical profession to practise fraud upon their patients in a manner that exposes and enacts an assumption of unrestricted ('supernatural') power, there is also something ridiculous and precarious about that power as it is staged in these theatricals. Cox tries to defuse the element of the 'ludicrous' (29) by insisting that in certain cases the patient's life is at stake. A man may die if he will not eat, from a conviction that his stomach is inhabited by 'a frog, snake, or toad, &c.' or that his digestive passages are blocked by a bone or a stone. In such cases, Cox advises, a doctor should be prepared to inflict a fraudulent superficial wound and feign successful surgery, showing the patient a bloody stone or bone as proof of its success. Alternatively, a physician might induce violent vomiting, secretly conveying 'a frog, snake, or toad, &c.' into the bowl (29–30).

The three most striking case-studies Cox provides to demonstrate pious fraud at work are all examples of hypochondria in men, and all are models of the efficacy of pious fraud. Mr A——, aged 25, fond of anatomy and animal dissection, 'had read some medical authors with much attention, and was in the constant habit of quacking himself' (33). His madness takes the form of a conviction that he has contracted syphilis 'not by any unfortunate connection, but from sitting on the same seat with an infected person' (34). Cox cures him by arranging for a fellow physician, whom the man has long admired, to send a fake prescription. 'Mr ——', aged 36, is a man much attached to literary pursuits, committed to Cox's care when his tendency to depression takes an unprecedented turn to paranoia. He becomes convinced that his housekeeper intends 'to destroy him by means of a succession of poisoned shirts'. Cox effects a cure by having the housekeeper served with a pretended warrant for her arrest and carried away, 'notwithstanding all her protestations of innocence', then treating her master with harmless 'antidotes' (*Practical Observations* becomes noticeably vague at this point but gives the impression that the housekeeper is not informed of the plan, in order to ensure that her performance will be convincing; 34–5). Next, the doctor tells of a patient whose excessive hypochondria is cured by passing 'a bougee up the urethra', informing the man that the healthy deposits of mucus from the glands are pus: evidence of 'universal ulceration' of the kidneys and bladder—which is then treated (37). Justified in his anxiety, the man is enabled to exorcise his fears, and is returned to society a well man.

As Cox proceeds with other and more varied medical histories, it becomes evident that these three cases of pious fraud, told at length and to great comic effect (not always deliberate), form a model of his psychiatric practice. What counts in every case is the demonstration of a specifically *medical* power in the absence of genuinely medical competence. In fact the means of cure are clearly not medical in any strict sense (surgery is faked; the pharmaceutics are panaceas), but the psychological effect justifies each deceit. As far as the patient is concerned, the doctor has admirably cured the body; as far as the reader is concerned, the doctor has given evidence of something quite different: superior force of mind. Cox was one of the leading proponents of 'moral management' therapy, according to which the physician's first and foremost resource was his moral command over the patient. The notion of 'pious fraud', even in the abstract, raises a number of doubts about how genuinely 'moral' the doctor's authority may be, but its more revealing aspect is that it ascertains at the start of Cox's 'practical' observations that what really counts is not medical expertise but the demonstration of power. Even where medical practice consists of the more orthodox administration of medicines (as it rarely does), the acknowledged foundation of psychological medicine is the production of the 'strong impression'. All medical intervention in cases of insanity can thus be seen as the pious fraud that screens the truth of the physician's practice: his ability to enforce his authority over the patient.

Cox's narratives are not all told with the same ease, however. His accounts of treating insanity in women differ sharply from the theatrical, often comic, but always successful narratives which record his treatment of male patients. To read Elaine Showalter or Philip Martin, it would be easy to assume that medical textbooks on insanity were predominantly concerned with the aetiology and treatment of women's maladies. This is not in fact the case, even at the end of the period studied in *Love's Madness*. Most general treatises on insanity seem to have had remarkably little to say on the subject, given the general preoccupation with hysteria in English culture, and the high proportion of asylum inmates who were female.[25] Cox presents only six cases of female insanity at any length, and in these *Practical Observations* markedly departs from the narrative structures just described.

Case VII, aged 19, has a temperament 'not very exquisitely marked, but rather choleric'. She is the only patient whose physical appearance as well as character is described; she has 'fair skin, dark hair and eyes'.

> A tender attachment to a worthless object at length diminished her natural vivacity, she became pensive, and fond of solitude. It was soon discovered that one of those accomplished villains, who so frequently practise their successful systems of seduction, after securing her confidence, had at length triumphed over her too susceptible heart, rioted in the possession of her charms, and then basely deserted her. I cannot pretend to pourtray what beggars description; but a more interesting or distressing case could scarcely be imagined; suffice it to say, the period of gestation was passed without any peculiar corporeal indisposition, but the unhappy patient pined in secret, her vivacity and spirits, like her deceitful lover, abandoned her, and her countenance exhibited the most striking traits of guilt and despondency. A protracted and painful parturition reduced her delicate frame to extreme debility and emaciation, while her ideas became confused, and her mind obviously diseased. (64)

'A more interesting or distressing case could scarcely be imagined': this is the only occasion upon which Cox expresses any particular interest, any emotional or imaginative attachment to a patient, and it is not difficult to comprehend why she should be the patient to call forth such a response. The 19-year-old's case is the stuff sentimentalism is made of. Cox tells us that his patient possesses 'very superior literary acquirements' (64), and her story is, in every detail, the same as those familiar from the fiction in which she is, presumably, well versed.

Cox's description of his patient seems at first to pick up the familiar sentimental narrative unproblematically. The subject-matter evidently attracted him, and he incorporated an extended reflection on the dangers of seduction into the opening pages of the

[25] See Mark S. Micale, 'Hysteria and its Historiography: The Future Perspective', *History of Psychiatry*, 1 (1990), 53–60, 93–101 for discussion; and cf. Showalter, *The Female Malady*, 52.

second edition.[26] Difficulties emerge, however, when he moves from description to treatment. While the sentimental story is a ready vehicle for Cox's stylistic exuberance—belying the protestation that he 'cannot pretend to pourtray what beggars description'—it is, nevertheless, a vehicle limited in its usefulness for him. This is a narrative which speaks only of the causation of sickness and not of its cure. The confinement of its focus to the moment at which health fails is underlined by the way in which the case-study reduplicates the narrative so that the lady's health repeats her lover's unkindness and abandons her. In describing the process of her cure, Cox is necessarily in the position of adding a somewhat awkward post-script to a narrative which, so far as his readers will be concerned, is already complete. Given that the basis of her story is absolutely familiar from late eighteenth-century sentimental fiction, there is imaginatively nowhere for it to go once past the point of her madness—which is the point from which Cox's role must begin. Imaginatively speaking (and Cox's recognition of the lady's imagi-native appeal has already been noted) there should be nothing more to say about her beyond the statement that she has fallen into insanity—unless she were to die, which is what the doctor, by definition, is there to prevent.

It is, therefore, doubly significant that Cox uses no more precise categorization of insanity than the phrase 'her mind was obviously diseased'. The unfortunate lady's cure is gently holistic, working through the restoration of the body to the restoration of the mind—a mode of treatment which differs markedly from Cox's moral management and his ready recourse to invasive practice in most other instances. The patient is plied with a simple diet, gradually enriched; medicines are administered; and 'the mind was kept inter-ested by change of scenery and varied pursuits: the sympathy of kind friends, and the consolations of religion, brightened her future prospects, and elevated her hopes. Under this system, both mind and body daily acquired strength, and at length health and reason were perfectly re-established' (65). Re-establishment is not a possi-bility generally encompassed by the sentimental narrative conven-tion which Cox invoked at the start of this patient history. It may be for that reason that no such romantic case occurs again. Cox's

---

[26] Joseph Mason Cox, *Practical Observations*, 2nd edn., corr. and greatly en-larged (London, 1806), 23–6.

next case history but one might, however, be read as a commentary upon the slightly uncomfortable status of such a supplement to, and cure of, the sentimental female malady.

Case IX is announced as an example of insanity arising from excessive physical strain, compounded by inadequate diet. It is cured by the same means as Case VII, and the parity of treatment is itself an invitation to read them in parallel. Case IX, however, is the most difficult of Cox's patient histories to interpret, because its tone is so problematic. 'Miss ——', aged 22, is very delicate, partial to botany and drawing, and to rambling all over the countryside in pursuit of those hobbies, regardless of the weather or of when she last ate. '. . . after a long ramble over rugged steeps, precipices, and mountains, in one of the most romantic parts of North Wales, she was fortunately met with by a peasant, when she exhibited all the usual symptoms of furious madness, seated on a hillock, surrounded by fragments of plants and drawings, using the most frantic gesticulations, uttering violent vociferations, and spouting parts of Shakspeare' (66–7). She is carried home, with great difficulty, where she is bled and purged to no good effect, at which point Cox is called in, and succeeds in curing her by the same methods he used earlier.

Here, pathos seems to have tipped over into parody. Miss —— is clearly cast in the role of Ophelia, surrounded by scattered—and apparently rather tattered—plants. Her spoutings of Shakespeare press the point home, but this is no fragile heroine: her marathon ramble 'over rugged steeps, precipices, and mountains' is bizarrely unlikely. And what are we to make of the peasant? The narrative sequence seems to imply a causal link between the appearance of the peasant and the outbreak of insanity, but this is presumably the result of loose wording on Cox's part. As he presents the case, the distinction between cause and symptom is very unclear: is a penchant for long walks the cause of insanity or is it a symptom that derangement has already set in? Perhaps most perplexingly, why the Ophelian touches? The Shakespearean allusion works parodically rather than sympathetically, turning the gathering of flowers into a botanical monomania rather than the means to the more coherently coded expression of feminine distress found in *Hamlet*.

Practical Observations seems to have a recurrent problem with tone when it deals with female patients. There are only three more discussed at any length: 'Mrs ——, aged 50 . . . an illiterate and naturally gloomy' woman, provides Cox with an opportunity for demonstrating the dangers of religious excess (39–41). Case XVIII—an altogether more difficult woman—proves to be the most recalcitrant patient under the swinging treatment. 'Miss ——', aged 25, delicate, nervous, and prey to a lung illness 'without any obvious cause became unusually talkative, sported some very strange ideas, and at length from confused thought became compleatly insane' (Cox notes that her lung disease is driven out by the insanity; 116–17). When other methods prove ineffective, Cox prescribes the swinging treatment, only to find that his patient enjoys it and that extraordinary levels of velocity and height are required to produce the dizziness, nausea, and fear which effect a cure. When Case XX looks set to be just as difficult, Cox will have none of it. The swinging treatment terrifies her, but he persists in the face of her pleadings until, at length, 'both health and reason [are] fully reinstated' (124–30).

Something goes wrong in *Practical Observations* when it introduces the subjects sentimental fiction would lead us to expect could be handled most securely. Far from there being a ready reciprocity between medicine and fiction, in Cox's 1804 text the most culturally established narratives about insanity do not work. Precisely because those narratives *are* established, *are* self-contained, *are* predetermined, they allow no space for the imposition of the wider narrative structure *Practical Observations* is concerned to follow: the narrative which leads from the derangement of the patient, through the physician's intervention, to the patient's cure.

John Conolly and the Alternative Uses of Literature

John Conolly's *An Inquiry Concerning the Indications of Insanity with Suggestions for the Better Protection and Care of the Insane* was published in 1830, twenty-six years after Cox's treatise. Conolly was then Professor of the Practice of Medicine at London University. He was also inspecting physician to the Lunatic Houses for the County of Warwick, and would go on to become the most famous alienist of his day. He, more than anyone, is credited with the reform of the English asylum system: the general acceptance of non-restraint and the practical institution of a more philanthropic and paternal approach to care of the insane.[27]

Though the *Inquiry* was a very early work for Conolly, Hunter and Macalpine recognize it as an important one in the history of psychiatry, for it was 'the first book which attempted to link normal and abnormal states of mind, "to render the recognition of insanity less difficult, by showing in what it differs from those varieties of mind which approach nearest to it"'.[28] It has also, as Andrew Scull points out, proved a more palatable book to twentieth-century readers than Conolly's later writing, which, despite a continuing opposition to restraint, increasingly advocates the institutionalization of the insane.[29] The *Inquiry* argues much more strongly for the freedom of all but the most dangerous among the mad. Though remembered for its libertarianism, however, the main part of Conolly's book, and the most intricately argued, is its attempt to theorize the nature of insanity. Like Cox, Conolly begins by stating his dissatisfaction with the current state of writing on insanity, but he goes on to argue more particularly that the medical profession has been at fault in its treatment of insanity as a negative phenomenon: that which is not sanity. Instead, it should be viewed as an affliction of the previously healthy body: a deviance from, not something in opposition to, the norm. The distinction serves a rhetorical purpose within the context of Conolly's desire for asylum reform. To perceive the insane as not inherently alien is an essential first step to the transformation of their care from a harshly corrective approach (such as Cox's book amply demonstrates) to a benevolently authoritarian one.

It would be easy to dismiss Conolly's *Inquiry*, on the grounds that its 'redefinitions' do nothing to clarify either deviance or insanity, and that the text succeeds only in reinstating the problems it claims to address. However strong its insistence on the continuum

[27] On Conolly's achievements, see Hunter and Macalpine, *Three Hundred Years of Psychiatry*, 805–9, 1030–8; and their introductions to the reprint editions of John Conolly, *An Inquiry* (London, 1964) and *On the Construction and Government of Lunatic Asylums* (London, 1968). Also Showalter, *The Female Malady*, 42–68. On the inaccuracy of attributing non-restraint to Conolly, see Andrew Scull, 'A Victorian Alienist: John Conolly, FRCP, DCL (1794–1866)', in Bynum *et al.* (eds.), *The Anatomy of Madness*, i. 124–5.

[28] Hunter and Macalpine, *Three Hundred Years of Psychiatry*, 806, quoting Conolly.

[29] Andrew Scull, 'A Brilliant Career? John Conolly and Victorian Psychiatry', *Victorian Studies*, 27 (1984), 203–35; and Scull, 'A Victorian Alienist', 103–50.

between sanity, peculiarity, and insanity, it has to resort to simple assertion when it finds madness, for it lacks the foundation in neuropathology necessary to ground its insights in medical science. When Conolly intersperses the treatise with a series of case-studies, they signally fail (as will be seen) to provide a workable model for the diagnosis and treatment of the insane. They do, however, work to enforce the general theoretical model by which Conolly claims to distinguish between the sane and the insane; and this broadening of the terms in which madness is being addressed is reflected in the case-study as a shift from Cox's flamboyant theatricals to a more sedate interest in the conceptual models offered by literature.

Unlike Cox, Conolly is more interested in discussion than demonstration, involving the physician in a philosophical rather than a curative or even a diagnostic role.[30] Where Cox provides histories of his own patients, Conolly takes most of his case-studies from other writers and earlier treatises on insanity. A striking number of them are literary—either indirectly, in that they involve celebrated authors (Ben Jonson, Dr Johnson, Cowper, Pascal, Lord Byron), or directly, in that they are taken from novels, poetry, and plays. Conolly is probably the best-known nineteenth-century medical reader of *Hamlet*—an interest which Charles Reade satirized viciously in his novelistic exposé of mid-century asylum practices, *Hard Cash* (1863). Reade's asylum director, Dr Wycherley (Conolly), reveals his own obsessiveness as his assertions of Hamlet's madness push him into monomania.

There is, however, little mention of Hamlet in the *Inquiry*, and there are no Ophelian, love-melancholic women among Conolly's case-studies. His references to female patients noticeably depart from literary stereotypes, and their tone tends to be oddly comic or strained. There are significantly few of them.[31] The one example of

---

[30] See, for example, his presentation of a patient suffering from megalomania (289–91). Andrew Scull has also noted Conolly's lack of faith in contemporary diagnostic procedures and medicines, and the hindrance his lack of practical flair posed at several points in his career as a physician. Scull, 'A Victorian Alienist', 106, 114–15.

[31] See esp. his mad schoolmistress, who mistakes her tables and chairs for pupils and beats them by night (*Inquiry*, 330–1); a woman who, after childbirth, suffers delusions of having killed her husband and is cured by his solicitous proof to her that he still lives (ibid. 402–3); and a letter quoted from a female patient of Dr Reid, describing a recurrent terror that she will set fire to the house (the case is dismissed with the declaration that she is 'altogether a lunatic' when in the grip of these delusional spells; *Inquiry*, 335).

love-madness discussed at any length involves an 'unfortunate gentleman' who 'fancies that a princess is in love with him' and 'wanders about the woods . . . carv[ing] the name of his beloved on trees' and 'indit[ing] moving letters to her in cherry juice' (384–5). His commital to a lunatic asylum by relatives eager to seize his money strikes Conolly as a grotesque injustice (386). The *Inquiry*'s representative figures from literature are also male, with very few exceptions. The example upon which Conolly hinges his discussion of madness as unmoderated self-examination is that of Macbeth:

Of this form of madness Shakspeare has given us a remarkable illustration, which, although that immortal poet has frequently been referred to by writers on the subject of mental disorder, has not, I think, attracted attention. Yet it is one of the most accurate, and one of the most complete. It is presented in the character of Macbeth; in which, as in that of Hamlet, Shakspeare seems to have meant to delineate the influence of strong emotions, although of a very different kind, on minds unequal to bear them. In both cases, disorder of the mind is produced, but the species of disorder is very different in the two cases. (319)

A three-page analysis of the play follows, the crux being that Macbeth's disorder of mind is more genuinely authenticated by the text than Hamlet's, and that he is seen to triumph over delusion by comparing and reasoning against reality:

> . . . . I see the[e] still;
> And on thy blade and dudgeon, gouts of blood,
> Which was not so before.—*There's no such thing*:
> It is the bloody business which informs
> Thus to mine eyes.

(321–2; Conolly's emphasis)

The significance of Conolly's shift into literary criticism is that for him the literary reference should have the same authority as the authentic case. This is 'one of the most accurate, and one of the most complete' illustrations available of the mad/not-mad distinction. In part, as Hunter and Macalpine have suggested, the predominance of literary example is probably a consequence of Conolly having, in 1830, limited clinical experience to draw upon.[32] Nevertheless, literature facilitates his argument quite as

[32] Hunter and Macalpine, Introduction to Conolly, *Inquiry*, 2.

readily as his few first-hand case notes. Where Cox found literature's madwomen a hindrance to medicine, Conolly finds its not-
quite-mad men an aid. The difference is partly in the gendered
construction of the literary models appealed to: Cox struggles with
a closed narrative of feminine madness which aestheticizes women's sickness and does not foresee their cure, whereas Conolly
benefits from less fixed, less comfortably sentimentalized, and
therefore more readily questionable representations of male insanity. But it is, equally, a difference in the kind of medical writing
involved. Cox looks for an escape from the vagaries of generali-
zation in the demonstrable power of the doctor, and finds a particular literary vogue (the sentimentalized madwoman) unhelpful;
Conolly looks for an escape from uninformed medical practice in
the formation of a general theory, and finds other kinds of literary
allusion very helpful indeed. Shakespeare's play offers a conceptual
coherence and a model of verbal reasoning uncomplicated by the
presence of the patient. Macbeth's illness, in other words, is available for Conolly's reading as a purely theoretical encounter with
delusion. Only in such a form is Conolly's theory finally workable.
The *Inquiry* begins with the question of the patient—to restrain or
not to restrain[33]—but it achieves its answer finally by setting aside
the patient and addressing the definition of insanity through the
controlled medium of literature.

In Conolly's writing, elegant reference to literature thus provides
the terms for a theoretical distinction between sanity and insanity in
the absence of practical means, but it simultaneously serves a more
simple and crucial function. It signals that the physician is a well-
educated gentleman, a man of polite learning. Conolly's representative figures are not only men of literary prominence but also men of
social consequence: he draws on memoirs or recollections of Henry
III of France, the Duke of Schomberg, Nicolai of Berlin, Sir Philip
Sidney, Sir Isaac Newton . . . the list goes on. The class-cultural
codes of Conolly's rhetoric gain considerably in significance once
the difference between Cox's and Conolly's status within the medical profession, and more broadly within society, is taken into
account. Cox's position as a gentleman physician in Gloucester-
shire was secure in 1804. He had inherited a practice established
fifty years before by his grandfather, and was to pass it on to his son
(it descended to succeeding members of the family until 1852).

---

[33] The reference in this instance is to incarceration rather than to strait-jacketing.

Conolly's background and his place within the profession were not so clearly respectable. As Andrew Scull has described it, Conolly's early life was at best shabby-genteel, and his career in medicine, while it brought him the social status he desired, was accompanied by a double-edged concern to maintain that status: medicine secured him the character of a gentleman but he equally had to maintain his character as a gentleman if he was to be a respectable doctor. A strong commitment to working men's education in no way detracted from Conolly's sensitivity to his own social position.

When the *Inquiry* appeared, Conolly had, to his own surprise, escaped an unsuccessful practice in the provinces to take up the position of Professor of Medicine at the new University of London, but there was no guarantee that the rise in his professional fortunes would continue. Relations with the university administration and with his students were becoming increasingly difficult in 1829–30, and in early December 1830 Conolly resigned his post.[34] In reviewing the affair, Adrian Desmond characteristically underlines the element of class-anxiety in Conolly's action. He also usefully resituates Conolly within a heated debate over the control of medical knowledge in this period, fought out largely in relation to competing religious and evolutionist views of physiology. Conolly deeply resisted the incursions being made on the medical profession by a new breed of materialist science, frankly mechanistic in its outlook on human life and intent on levelling the old hierarchical structures of the medical profession. He was passionate about the danger of students being attracted by the 'scientific pyrotechnics' of men like William Lawrence, author of the highly controversial *Lectures on Comparative Anatomy* (1819). In his own classes, Conolly did his utmost to impress upon his students that 'not *knowledge* alone, but *character* is power . . . knowledge without character can procure no more than temporary and very transient pre-eminence'[35] (the language is strongly suggestive of his growing interest in phrenology).[36] These issues were very much in the foreground at the time he was writing his own *Inquiry*. In 1829

[34] See Hunter and Macalpine, Introduction to Conolly, *Inquiry*, 28–32; Scull, 'A Victorian Alienist', 112–18.

[35] Quoted in Adrian Desmond, *The Politics of Evolution: Morphology, Medicine, and Reform in Radical London* (Chicago, 1989), 204.

[36] On Conolly's advocacy of phrenology, see Roger Cooter, *The Cultural Meaning of Popular Science: Phrenology and the Organisation of Consent in Nineteenth Century Britain* (Cambridge, 1984), 32, 93, and, particularly, 331 (n. 71).

Southwood Smith's radical materialist *Animal Physiology* had gone to the publishers and Conolly, one of the book's referees for the Society for the Diffusion of Useful Knowledge, was calling loudly for its substantial revision or retraction. As Desmond argues, the debate over the content of medical physiology concealed profound anxieties over the future of the gentlemanly physicians—the medical élite whom Conolly had so long aspired to join.[37] The interfusion of the literary and the medical in Conolly's *Inquiry* needs to be read, in part at least, as a response to those concerns. Certainly it would become a far more dominant factor in his later writing.

In light of the conviction and style with which Conolly proposed his libertarian ideas in 1830, the arguments pursued in his later works are, to say the least, surprising. An account of the non-restraint methods practised at Hanwell appeared in 1847,[38] and in 1856 he published a book which, from its title alone, would seem to follow naturally from the earlier works: a defence of *The Treatment of the Insane without Mechanical Restraints*. In the interim, however, Conolly's position had changed considerably. Success had not brought riches. Financial circumstances had obliged him to open a private asylum for female patients and to make what money he could from acting as an expert witness on insanity in criminal trials and cases involving disputed grounds for committal to an asylum. He therefore found himself in the uncomfortable position of having to argue against his earlier libertarianism. Once again, however, literature proved his ally. In a remarkable passage, which Elaine Showalter uses to powerful effect in *The Female Malady*, Conolly imported the figure of Bertha Mason from Charlotte Brontë's *Jane Eyre* into his argument for not keeping lady patients in their own homes. They too easily become 'quite estranged' from all their relatives, he warned his readers. Their conduct becomes 'fierce and unnatural', and the house itself is

[rendered] awful by the presence of a deranged creature under the same roof: her voice; her sudden and violent efforts to destroy things or persons; her vehement rushings to fire and window; her very tread and stamp in her dark and disordered and remote chamber, have seemed to penetrate the

---

[37] Desmond, *Politics of Evolution*, 204.
[38] John Conolly, *On the Construction and Government of Lunatic Asylums and Hospitals for the Insane* (London, 1847).

whole house; and, assailed by her wild energy, the very walls and roof have appeared unsafe, and capable of partial demolition.[39]

Bertha Mason was ideal material. Utterly divorced from the sentimental convention of love-madness, she was a grotesquely unarguable case for professional intervention. Moreover, the extraordinary success of *Jane Eyre* was the perfect social cover, enabling Conolly to conduct his about-face with some flair.

Only in 1863 did Conolly produce the work on *Hamlet* for which he has most often been remembered by literary critics. *A Study of Hamlet* was not the only medical interpretation of Shakespeare's plays to appear in the period. J. C. Bucknill's two-volume study *The Psychology of Shakespeare* and *The Medical Knowledge of Shakespeare* had appeared in 1859 and 1860, but Conolly rapidly became the more celebrated medical reader of *Hamlet*, and the Shakespearean touch was peculiarly identified with him. Both Conolly and Bucknill were probably responding in part to a contemporary resurgence of artistic interest in Ophelia. John Everett Millais's Pre-Raphaelite painting of Ophelia's drowned body and Arthur Hughes's more sentimental depiction of her seated on the willow-bough were both exhibited at the Royal Academy in 1852. They were only the most lastingly successful examples of a much larger Ophelian industry. At least fifteen paintings of her were exhibited at the Royal Academy alone in the first half of the nineteenth century.[40]

The *Study of Hamlet* is a much slighter volume than any Conolly had published previously, and it is also the most flamboyantly self-indulgent. Elaine Showalter makes forceful use of his views on Ophelia as further evidence for the influence of literary and aesthetic models of femininity on Victorian medical practice.

In his *Study of Hamlet* in 1863 he noted that even casual visitors to mental institutions could recognize an Ophelia in the wards: 'The same faded

---

[39] John Conolly, *The Treatment of the Insane without Mechanical Restraints* (London, 1856), 149–50; quoted in Showalter, *The Female Malady*, 68.

[40] Richard D. Altick, *Paintings from Books: Art and Literature in Britain 1760–1900* (Columbus, Oh., 1985), 299–302, and pls. 78, 215, 216, 217. It is also worth noting the interest in Ophelia sculptures—a subject usually neglected. John Graham Lough's life-size marble 'Ophelia' (1841), exhibited at the Royal Academy in 1843, attracted considerable debate among art critics. He returned to her as part of his Shakespearean frieze, executed for the Ridley family in 1855. See John Lough and Elizabeth Merson, *John Graham Lough 1788–1876: A Northampton Sculptor* (Woodbridge, 1987), 58, 84.

beauty, the same fantastic dress and interrupted song.' . . . Conolly urged actresses playing Ophelia to come to the asylum and study real madwomen. 'It seems to be supposed,' he protested, 'that it is an easy task to play the part of a crazy girl, and that it is chiefly composed of singing and prettiness. The habitual courtesy, the partial rudeness of mental disorder, are things to be witnessed. . . . An actress, ambitious of something beyond cold imitation, might find the contemplation of such cases a not unprofitable study.'[41]

But what was the context of these remarks, and how far were they typical expressions of medical opinion and medical practice in their time? In fact, the *Study of Hamlet* was the only one of Conolly's works not to be published by a medical or academic press. All his earlier writings had been published by John Taylor (the University of London's official publisher) or by J. J. Churchill & Son, London's leading publisher of medical texts, but the *Study of Hamlet* appeared with Edward Moxon. Best known as Tennyson's publisher, Moxon also handled the work of Wordsworth, Lamb, Shelley, Martineau, Patmore, and Browning, amongst others. Although Moxon extended his interests beyond the purely literary in later life, he remained primarily a literary publisher and a familiar presence in fashionable intellectual society. The very circumstances of the *Study of Hamlet*'s publication therefore mark it out as being not in the same field as Conolly's more substantial works. Moreover, the appearance of the book three years after Bucknill's was probably not accidental. Bucknill was at the head of the alienists' profession in the early 1860s: he was appointed President of the Association of Medical Officers of Asylums and Hospitals for the Insane in 1860, and two years later he became Lord Chancellor's Visitor in Lunacy. Conolly was at the very end of his professional career, and unlikely to be vying for professional advancement. It is far more likely that he saw Bucknill's book as treading on ground in which he had a long-standing personal interest.[42]

The purpose of the *Study of Hamlet* was, in other words, very different from that of Conolly's *Inquiry* or the treatise *On the Construction and Government of Lunatic Asylums*. This was not primarily a medical work, but one directed towards a lay audience,

---

[41] Showalter, *The Female Malady*, 90.

[42] On Conolly's involvement with the Shakespeare Library, the Shakespearean Club, and, particularly, the Shakespeare Monumental Committee at Stratford, see Hunter and Macalpine, *Three Hundred Years of Psychiatry*, 13.

aiming to capture the attention of fashionable society. Contrary to widespread assumption, Ophelia was not the prototype of female insanity for nineteenth-century alienists. By mid-century, she rarely entered the academic writing of even the most urbane physicians; instead she belonged increasingly to a distinctly secondary body of writing whose principal significance is as a reflection on the social status—and the literary pretensions—of the authors. Notably absent from Conolly's medical writing, Ophelia came into her own in a context where medical treatment was no longer the main issue. Conolly's remarks on the 'faded beauty, fantastic dress and interrupted song' of an Ophelia on the wards are not primarily the observations of a clinical eye: they are first and foremost evidence of a feeling heart and a refined sensibility—those eminently desirable qualities of the fashionable Victorian doctor.

Throwing Physic to the Dogs: Henry Maudsley and Sir Alexander Morison

The opposition to perceived incursions upon gentlemanly medicine for the insane discernible in Conolly's writing is even more evident in the writing of his son-in-law, Henry Maudsley, hailed in his own time and since as the leading alienist of the later Victorian period.[43] *The Physiology and Pathology of the Mind*, written during the early 1860s and first published in 1867, is Maudsley's best-known book, and the one he returned to and revised most often among his many publications on insanity. Maudsley's text inherits unchanged many of the preoccupations of Cox and Conolly. In one of the clearest echoes, it begins with an epigraph from Goethe warning against the dangers of speculation. It also echoes Conolly's concern with the absence of a clinical pathology of insanity which could ground medical philosophy firmly in physiology, and it reveals the same anxiety found in Cox's writing about maintaining the difference between practitioner and patient: 'if a looker-on could appreciate the force of the reasoning by which such a delusion was generated,' writes Maudsley, 'he would be as mad as the patient' (325).

[43] On Maudsley's career, see Trevor Turner, 'Henry Maudsley: Psychiatrist, Philosopher and Entrepreneur', in Bynum *et al.* (eds.), *The Anatomy of Madness*, iii. 151–89.

The (presumably conscious) allusion to Polonius—'to define true madness, What is't but to be nothing else but mad?'—encapsulates a semi-suicidal quality recognizable in much of Maudsley's rhetoric here and in later books. However much *The Physiology and Pathology of the Mind* shares with the earlier texts considered here, it also reveals a new set of problems in writing about the mad, particularly in its engagement with evolutionary theory. Maudsley's commitment to the future of medical science encourages him to read both science and society according to a model of evolutionary advance, yet his subject (and the means through which he hopes to establish medical science) is, by his definition, that which runs counter to and threatens to stall evolutionary progress. Indebted more to Spencer than to Darwin,[44] Maudsley seems almost willingly to embrace a concept of evolution unamenable to a therapeutic medicine for the insane. Viewing degeneration as the waste process of evolution, pathology as the refuse of physiology, Maudsley is left with an unadmitted but fundamental doubt about the role of the doctor in such a system. Why should the doctor seek to preserve, or to restore to society, lives which nature has marked unfit for assimilation? Significantly, Maudsley's text ends by abandoning practical medicine and with it the smaller narratives which, for Cox and Conolly, define and legitimize it: the drive for a cure in the individual case. *The Physiology and Pathology of the Mind*'s closing pages advocate emptying the asylums of all but their most dangerous patients, at least until such time as 'microscopical' chemistry[45] will unravel the secrets of the physiology of mind (the fact that Maudsley's libertarianism is far stronger than Conolly's here has remained surprisingly unremarked).

Like Conolly's *Inquiry*, *The Physiology and Pathology of the Mind* finds in literature a powerful resource against the inadequacies of contemporary medical science and its rhetoric. Maudsley's writing is remarkable for its use of quotation from a wide range of European literature, as Aubrey Lewis made clear in his 1951 Maudsley Lecture:

---

[44] Maudsley cites Spencer in his preface, though Spencer himself complained that the debt was insufficiently acknowledged. See Turner, 'Henry Maudsley', 186 (n. 130).

[45] Henry Maudsley, *The Physiology and Pathology of the Mind* (London, 1867), 57. See also pp. 65, 178, 369, 375.

The writers whom he most often quotes are significant: Goethe perhaps more than any other; Shakespeare, the Bible (especially Ecclesiastes and Isaiah), Milton, Bacon, Pascal, Montaigne, Robert Burton, Sir Thomas Browne, Gibbon, Amiel; and of philosophers, Locke, Hobbes, Hume, Hamilton. These are lofty minds, often of a sombre and sceptical cast, like his own. Then there are the medical and scientific authors with whom he was evidently familiar: Darwin, Huxley, Carpenter, Laycock, Prichard, Brown-Séquard (but he never quotes Galton, and seldom Hughlings Jackson, though he and Jackson were both close to the thought of Herbert Spencer . . .). With the German and French writers of his middle life he was as familiar as with his English contemporaries; but he seems not to have kept up his psychiatric reading after the eighties and nineties as much as his general reading (which included Karl Marx's writings).[46]

The list indicates an impressively well-read man; but though Lewis insists that the names catalogued are 'significant', he fails to say why. Maudsley's penchant for Goethe might be explained partly by reference to Goethe's emblematic status as a man who could cross fields between literature, science, and political administration,[47] but, for much of the time, Maudsley's allusions to literature seem largely decorative, and once again the decorativeness possesses both an immediate rhetorical function and a wider class-cultural one.

Maudsley's appeals to literature involve a conscious valuation of the creative, and not simply the representative, functions of art. He will end a chapter with an extensive quotation from *Paradise Lost*, use a passage from Goethe's *Faust* for an epigraph, cite *The Winter's Tale* as a footnote to a discussion of the scientific imagination, and so on, because in each case the literary sentiment confirms, and gives lyric utterance to, the scientific insight. In their use of quotation, nineteenth-century scientific writers were not much concerned with literary context. Maudsley has little interest in where a passage appears in *Faust*, or how *The Winter's Tale* progresses, or what

---

[46] Lewis's lecture was delivered in 1951; repr. as the introduction to Henry Maudsley, *The Pathology of the Mind* (London, 1979), unpaginated. The most extended discussion of Maudsley's literary influence to date is to be found in Michael Collie, *Henry Maudsley, Victorian Psychiatrist: A Bibliographical Study* (Winchester, 1988), 57–68.

[47] Maudsley shares with Goethe a strong belief in the moral and political ordering of nature. See Miles W. Jackson, *Goethe's Law and Order: Art and Nature in 'Elective Affinities'*, doctoral dissertation, University of Cambridge, 1991. Maudsley cites Goethe's scientific writing only once here (*Physiology and Pathology of the Mind*, 55), but quotes frequently from the poetical works.

concept of epic is involved in *Paradise Lost*. More simply and fundamentally, literary allusion represents for Maudsley the ideal of powerful utterance *per se*: the possibility of giving comprehensive imaginative expression to the natural world. His discussion of the nature of imagination near the end of 'The Physiology of Mind' leads into an assertion of that ideal:

As organic growth and development take place in obedience to the laws of nature, and yet constitute an advance upon them, so it is with the well-cultivated or truly developed imagination, which brings together images from different regions of nature, yokes them together by means of their occult but real relations, and, thus making the whole one image, gives a unity to variety: there is an obedient recognition of nature, and there is a developmental advance upon it. This *esemplastic* faculty, as Coleridge, following Schelling, named it, is indicated by the German word for imagination, namely, *Einbildung*, or the *one-making faculty*. Its highest working in our great poets and philosophers really affords us an example of creation going steadily on as a natural process; and creative or productive activity is assuredly the expression of the highest mental action . . . (187–8)[48]

Numerous oppositions are held together in this passage: the real and the occult, the natural and the cultivated, evolutionary creation and human creativity, the law and the imagination, even (linguistically) German and English. Maudsley at once acknowledges and denies their boundaries, insisting upon the cohesive force and, indeed, the curative power, of the creative imagination, very much as Tyndall was to do in the 1870s.[49] By such an investment in the power of literature, Maudsley's text establishes, still more strongly than does Conolly's, a resource which takes it beyond the limitations of contemporary medical science. In making literary allusion part of the fabric of its writing, *The Physiology and Pathology of the Mind* constantly invokes a principle of conceptual cohesion which contemporary medical science could not offer, at least not on terms Maudsley was prepared to accept.

If literature speaks to the limitations of contemporary medical knowledge, it also speaks to what, for Maudsley, were its less attractive advances. The following passage comes from the beginning of part II, and its place in the text confirms its relationship to

---

[48] See also Maudsley, *Physiology and Pathology of the Mind*, 66, 185.

[49] John Tyndall, *Essays on the Use and Limit of the Imagination in Science* (London, 1870), esp. p. 16 on *Einbildungskraft*.

the problems of yoking 'The Physiology of Mind' with 'The Pathology of Mind':

Do we not, in sober truth, learn more of [insanity's] real causation from a tragedy like 'Lear' than from all that has yet been written thereupon in the guise of science? An artist like Shakespeare, penetrating with subtle insight the character of the individual, and the relations between him and his circumstances, discerning the order which there is amidst so much apparent disorder, and revealing the necessary mode of the evolution of the events of life, furnishes, in the work of his creative art, more valuable information than can be obtained from the vague and general statements with which science, in its present defective state, is constrained to content itself. (198)

Maudsley's 'sober truth' overturns any privileging of science over fiction. Great literature, by his argument, surpasses science in that it orders the disordered appearance of insanity, relates insanity back to the development and structure of society, tells its story as evolution, while 'all that has been written in the guise of science' can speak only of devolution and degeneration. Science, not insanity, is the defective and disordered subject—but defective and disordered in a specific as well as a general sense. The language with which Shakespeare is enlisted above offers one of the few glimpses of a quarrel with developments in the contemporary medical profession otherwise hidden behind the flowery pessimism of *The Physiology and Pathology of the Mind*. It puts Maudsley in this respect very close to his father-in-law, despite their differences over the future of medical science.[50] As Trevor Turner has shown, Maudsley was among the staunchest defenders of the gentility of his profession. He fought to maintain the social prestige of the profession, initially within the Association of Medical Officers of Asylums and Hospitals for the Insane, but from 1865, when it became the Medico-Psychological Association, he found himself increasingly in conflict with his colleagues. In 1865, as Maudsley was finishing the writing of *The Physiology and Pathology of the Mind*, he was also deeply embroiled in attempting to prevent the admission of laymen to the Association.[51] Maudsley's insistence on

[50] See Scull, 'A Victorian Alienist', 134.

[51] On Maudsley's deteriorating relations with the Medico-Psychological Association, see Turner, 'Henry Maudsley', 156. Turner sees Maudsley's movement away from the Association in terms of a personal intellectual shift that was out of key with the rest of the medical establishment, whereas I am seeking to place that shift in the context of 19th-cent. gentlemanly medicine as a whole.

the artist's 'penetration by subtle insight' thus sums up a conservative desire to safeguard the gentlemanly character of physicianship from the inroads of (largely middle-class) technicians. Hence, also, his forward gaze to the insights of microscopical chemistry. The alternative of less 'subtle' insights through hard technology is notably absent from his rhetoric.[52] Literary reference for Maudsley, as for Conolly, is more than a gentlemanly flourish: it reflects a desire to *preserve* medicine as a gentlemanly hermeneutic art at a time when gentlemanly medicine was under threat.

With so much emphasis on literature, Maudsley might have been expected to turn to the love-mad woman with rather more interest than Conolly. However, to move from Joseph Mason Cox to Conolly to Maudsley is to see the sentimental figure of the love-mad woman gradually relegated to a more and more marginal place, until, with this, the most self-displayingly literary of nineteenth-century physicians to the insane, she disappears almost entirely. Only one of his case-studies could be categorized as an example of love-madness, but it is not the term he chooses. Included in the selection of brief case notes appended to the chapter on cure is Case 11: 'A young lady, aet. 25, who had some anxieties at home, suffered a disappointment of her affections. Black depression running into acute dementia.—Recovery' (254). Why should the extensive literature on love-madness not have been a ready point of reference for Maudsley? Its omission seems the more remarkable in light of the mid-century resurgence of artistic interest in Ophelia. One of the reasons has already been suggested: hers was not a narrative which could readily be accommodated within a curative approach to insanity. But Maudsley's scepticism about the desirability of cure might have given him cause to appreciate the Ophelian model all the more. The more complex reasons for her exclusion from the writing of so belletristic a physician as Maudsley

---

[52] On opposition among senior British physicians to the introduction of instrumental aids during the 1860s and after, see Christopher Lawrence, 'Incommunicable Knowledge: Science, Technology and the Clinical Art in Britain 1850–1914', *Journal of Contemporary History*, 20 (1985), 503–20 (esp. pp. 504–5, 512). Opposition to technology did not detract from Maudsley's conviction that a better understanding of insanity would be achieved through a somatic, not a psychological, approach (there being a crucial distinction between technology and chemistry). See Michael J. Clark, 'The Rejection of Psychological Approaches to Mental Disorder in Late Nineteenth-Century British Psychiatry', in Scull (ed.), *Madhouses, Mad-Doctors, and Madmen*, 271–2.

can perhaps best be illustrated by turning back to the figure of Sir Alexander Morison.

Morison continued to be interested in love-madness when it had all but disappeared as a serious subject for other alienists, and that interest can be largely attributed to his age. When the fifth edition of his *Lectures* appeared in 1856, Morison was in his late seventies. His career very nearly spans the writings considered here. He was already 25 and practising medicine in Edinburgh when Cox's *Practical Observations* appeared, and he died just one year before the publication of Maudsley's *The Physiology and Pathology of the Mind*. Like Cox, Conolly, and Maudsley, Morison can claim a pioneering place in the history of medical specialization in mental diseases. His lectures, delivered in Edinburgh and London annually from 1823, were the first series of classes on mental pathology to be made available to students of medicine on a regular basis, and he was instrumental in the establishment of the first professorships in psychological medicine.[53] Like Conolly, Morison specialized for part of his career in the treatment of women, and like both Conolly and Maudsley he cultivated an extensive private clientele, including members of European aristocracy. (The knighthood conferred on Morison in 1838 was a mark of gratitude for his attendance on Princess Charlotte.[54] He was only the third doctor to the insane to receive such an honour in England.)

The fact that Morison grew up during the height of the sentimental cult may well explain his lasting interest in love-madness, but the anti-sentimentalism evident in those parts of the text quoted at the beginning of this chapter is typical of him, and it accompanies a more general determination to question received assumptions about the role of sex in relation to madness.[55] Morison's fusion of late eighteenth-century interest in love-madness with the increasing scepticism of the nineteenth century is particularly marked in the final section of his last works, which he dedicated to physiognomical portraits and accompanying studies of patients—and it

[53] See Hunter and Macalpine, *Three Hundred Years of Psychiatry*, 769.
[54] Morison dedicated *The Physiognomy of Mental Diseases* (London, 1840) to her husband Leopold, King of the Belgians, whose mental afflictions are alluded to in Charlotte Brontë's *Villette*.
[55] See particularly Morison, *Outlines of Lectures on the Nature ... of Insanity* (1848), 288–93.

is here that the reason for the gradual disappearance of love-madness from the writing of younger men can be most clearly discerned.

In the 1820s, as resident superintendent in charge of the women's wards at the Surrey County Asylum, Morison regularly invited artists to draw the faces of his male and female patients. It was a practice he continued when he became consulting physician to Bethlehem Hospital. Following in the footsteps of his mentor, Esquirol, Morison used the portraits as the basis for detailed examinations of the way in which insanity expressed itself in the shape and the muscular disposition of the face,[56] and the resulting case-studies accompany all the editions of his lectures from 1826 onward. Among the patients presented in the 1848 and 1856 editions were two women suffering from insanity brought on by a disappointment in love 'Eliza V——' and 'Couser D——'.[57]

'Eliza V——' was an unmarried lady's maid, aged 43. She was admitted to Bethlehem Hospital on 27 March 1846, 'labouring under an attack of Mania, complicated with Hysteria'. The causes of the malady were 'stated to have been disappointment in love, and erroneous views on religious subjects'. Morison reported that her behaviour had been sober enough until shortly before the mania made its appearance. Thereafter she was liable to become 'very spiteful and unmanageable. Her memory was impaired; she was subject to violent fits of Hysteria, and expressed herself in a very loose and sometimes incoherent manner. Purgatives and tonic medicines were employed . . . and she was discharged cured, after about two months treatment.'[58]

This is not the stuff of popular literature. Lacking Cox's stylistic pretensions, Morison's case notes are more carefully clinical. The possibility that disappointment in love was the exciting cause of insanity is stated tersely and seems to represent the view of those who committed her to the asylum, rather than one Morison himself is fully prepared to endorse. Any pathos the term 'disappointment

[56] A few of the illustrations accompany the *Outlines of Lectures* from the second edition of 1826 onward. An extended selection is attached to the fourth and fifth editions. By far the most expansive use of them is to be found in Morison, *Physiognomy of Mental Diseases*.

[57] The 1848 edition also gave details and a portrait of a male case. Morison, *Outlines of Lectures on the Nature . . . of Insanity*, 110.

[58] Ibid. 463; repr. in Morison, *Lectures on Insanity*, 5th edn., ed. Thomas Coutts Morison (Edinburgh, 1856), 463.

N.º 4.

FIG. 1 Plate 4 from Sir Alexander Morison, *Lectures on Insanity*, 5th edn. (1856)

FIG. 2 Plate 7 from Sir Alexander Morison, *Lectures on Insanity*, 5th edn. (1856)

N.º 8.

FIG. 3 Plate 8 from Sir Alexander Morison, *Lectures on Insanity*, 5th edn.
(1856)

in love' might convey is severely blunted by its being coupled with 'religious unorthodoxy'. A heavy-looking woman, with loose tangled hair, Eliza V—— is pictured seated on the floor, hugging her knees. She does not look like any known representation of Ophelia, and, though her station puts her closer to Cowper's serving maid, her behaviour evidently had none of the charm of Crazy Kate's. Moreover, unlike most literary madwomen, this woman was cured. Purgatives, tonics, and two months in Bethlehem Hospital saw her discharged—whether fit for service or not, Morison does not say.

The second case is equally unlike the models of female insanity offered by literary tradition. Figs. 2 and 3 depict a 23-year-old woman named 'Couser D——', before and after treatment in Bethlehem Hospital. Admitted a month before Eliza V——, this woman's disappointment in love was again not the sole cause of madness. It brought to the surface a strong hereditary strain of insanity (Morison records that both her father and her grandfather died insane). After a month in Bethlehem she appeared to have regained rationality, but, like any mid-nineteenth-century doctor, Morison laid great weight on that key indicator of female indisposition: irregularity of the menstrual cycle.

The menses . . . had not made their appearance; and she suddenly relapsed, and continued insane for about a month, when she again became quiet and rational. In this way, she relapsed three different times. Before her final recovery, she had an attack of erysipelas in her face; and shortly afterwards the menstrual discharge made its appearance, and she appeared quite convalescent. She remained two months longer in the hospital, and was then discharged cured.[59]

Again, there is no attempt to elicit the reader's sympathy. The cause of the illness having been briefly noted, Morison confines his attention to the symptoms of continuing sickness and then returning health. Erysipelas on the face is noted without any of Cox's or Conolly's mannered delicacy of expression. Most importantly, Morison's detailed charting of his female patients' mental health against their menstrual cycles signals a profound shift from late eighteenth and early nineteenth-century mad-doctors' emphasis on the general state of the nerves. For Cox, in 1804, female derangement was understood as a disorder of the whole system, requiring

[59] Morison, *Outlines of Lectures on the Nature . . . of Insanity* (1848), 464–5.

either that the patient's general health be restored by rest, good diet, and kind society, or, more brutally, that an effective shock be delivered to body and mind. Morison's late writing reflects a model of women's mental health firmly grounded in the far less holistic concept of *reproductive* health.

The fears that both Conolly and Maudsley reveal about the new breed of doctor with more 'technique' than 'insight' have no clearer expression than in the midwifery debates of the first sixty years of the nineteenth century. Gynaecology had its immediate origins in eighteenth-century male midwifery.[60] Often used as a quick route from surgery to physic by doctors looking for social and professional preferment, midwifery early on roused the ire of physicians jealous of their own privileges. During the year in which *Practical Observations* was published (1804), the Royal College of Physicians began repealing the legislation which, since the early 1780s, had allowed the most distinguished midwives to hold a Physicians' licence in midwifery.[61] The midwives were consequently obliged to consolidate their influence in general practice, but the campaign against the Royal Colleges and for recognition and regular constitution continued (it found a strong supporter in the alienist George Man Burrows, whose involvement made him highly unpopular with many of his colleagues[62]). The 1850s and 1860s were crucial years in the success of the midwives' long campaign. Though the terms of the 1858 Medical Act disappointed the most zealous reformers by leaving the Royal Colleges intact, they did provide for the regulation of obstetrics and gynaecology. A year later the Obstetrical Society of London was founded. The physicians and surgeons bowed to the inevitable and grudgingly, from the 1850s, began to open their doors.

As usual, institutional reform lagged well behind actual circumstances. The admission of obstetricians and gynaecologists to the inner sanctum of the medical profession was a belated recognition—and in effect a forced one—of a fundamental professional and moral reorientation of the practice of medicine. Specialist medical provision for women's diseases in the 1860s was almost

---

[60] See Ornella Moscucci, *The Science of Woman: Gynaecology and Gender in England 1800–1929* (Cambridge, 1990), 57–80; Roy Porter and Dorothy Porter, *Patient's Progress: Doctors and Doctoring in Eighteenth-Century England* (Oxford, 1989), 181–3.
[61] Moscucci, *The Science of Woman*, 57, 55.
[62] Ibid. 61.

unrecognizable compared with the situation even thirty years before. The new women's hospitals, rapidly expanding in number and
in prestige from the 1840s, provided a strong institutional framework and unprecedented opportunities for gynaecological and ob-
stetric research, and they greatly assisted the crucial shift in
emphasis from 'the study of the "whole woman" to the treatment
of a localised pathology'.[63] Surgical gynaecology expanded with
extraordinary rapidity from the early 1860s, and it summed up
everything the physicians had always disliked about midwifery and
about surgery more generally: it was invasive, it was brutal, it was
morally distasteful, and it was certainly not genteel. For physicians
specializing in insanity, it was particularly threatening in its implications, for if a woman's mental health depended on the state of her
reproductive functions, the gynaecological surgeon, not the psychological physician, could clearly claim prior authority.

The ascendancy of gynaecology did have something to offer
traditional physicianship. Most obviously, it provided a more
coherent justification than ever before for the doctor's scientific
authority. But for men like Maudsley the detractions outweighed
the benefits. By 1860 the intellectual validity of the gynaecological
model was no longer in dispute. Maudsley himself never questioned
the emphasis on the reproductive organs as the major determinant
in women's mental health. Indeed, one of his most widely quoted
works—though in many respects his least characteristic—is the
1895 pamphlet on *Sex in Mind and Education*, in which he argued
against women's higher education on the grounds that intellectual
strain carried a high risk of infertility in women. But the shift
towards gynaecology goes a long way towards explaining the disappearance of love-madness from the otherwise broad repertoire of
literary examples in his medical writing. Maudsley, for all his
opposition to incursions on gentlemanly physicianship, was intellectually of his generation, and though he used the weapon of
literary cultivation for all it was worth in the fight against the new
breed of women's doctor, he never alluded to Ophelia. Far from
assisting his case, she might well have damaged it. Ophelia would
have had short shrift and rapid surgical correction from many of his
colleagues.

Sir Alexander Morison was of an older generation, and it is not

[63] Moscucci, *The Science of Woman*, 108.

surprising that, of the two, it should be he who kept the memory of a more benign form of medicine for women alive in his sceptical but persistent attention to love-madness. The title-page to his 1848 lectures boasts a line from Gaubius: 'The care of the human mind belongs to the physician,—it is the most noble branch of our office.' The final lines of the 1848 and 1856 editions' discussion of love, grief, and insanity appear to reverse that statement exactly. Placed in the context of profound changes in the medical profession, however, it becomes clear that they represent a defence of a kind of physicianship quickly disappearing in mid-Victorian England. Like Maudsley's lines nine years later on the superiority of Shakespeare's *King Lear* to 'all that has yet been written in the guise of science', Morison's parting words reclaim the authority of the gentlemanly physician in the act of relinquishing that authority to literature. Medicine, he suggests, has little right, even in 1856, to intercede in a case of madness from disappointed love:

In the treatment of the depressing passions, occasional relief may be obtained by attending to the corporeal symptoms; but much is not to be expected from medicine,—more is to be effected by the soothing tenderness and consolation of a judicious friend, with seasonable admonition. . . .
Shakspeare makes *Macbeth* say:—

> 'Canst thou not minister to a mind diseased,
> Pluck from the memory a rooted sorrow,
> Raze out the written troubles from the brain,
> And with some sweet, oblivious antidote,
> Cleanse the stuffed bosom of that perilous stuff
> That weighs upon the heart?
> ——————— Throw physic to the dogs!
> I'll none of it.'[64]

---

[64] Morison, *Outlines of Lectures on the Nature . . . of Insanity*, 464–5; repr. in Morison, *Lectures on Insanity*, 119.

3

Hyperbole and the Love-Mad Woman: George III, 'Rosa Matilda', and Jane Austen in 1811

IN May 1811 George III was suffering a renewed attack of the severe physical and mental breakdown which had periodically debilitated him since 1788. His doctors, seeking to divert their patient during one of his more lucid spells, invited him to choose the programme for a concert to be held on 'the Duke of Cambridge's night' at Windsor Castle. The King had long been a dedicated admirer of Handel, and selected a programme of arias from his operas and oratorios: 'This consisted of all the finest passages to be found in Handel, descriptive of madness and blindness; particularly those in the opera of Samson; there was one also upon madness from love, and the lamentation of Jephthah upon the loss of his daughter; and it closed with "God Save the King" to make sure the application of all that went before.' It was, as the MP Francis Horner noted in a letter to his father, 'very affecting proof of [the King's] melancholy state' and a 'singular instance of sensibility; that in the intervals of reason he should dwell upon the worst circumstances of his situation, and have a sort of indulgence in soliciting the public sympathy'.[1] This would be one of the last such intervals of sanity he would enjoy.

George III furnishes the historian of madness with the most potent symbol of, and justification for, the late eighteenth and early nineteenth centuries' concern with madness. The strait-jacketed mad monarch signified 'constitutional crisis' of a public as well as private kind: a double threat which found ready expression in a language resonant with metaphors of the body politic.[2] European

---

[1] Francis Horner to his father, 16 May 1811. *The Horner Papers: Selections from the Letters & Miscellaneous Writings of Francis Horner, M.P. 1795–1817*, ed. Kenneth Bourne and William Banks Taylor (Edinburgh, 1994), 689.

[2] As Allan Ingram has shown, it was a culture which found the contemplation of insanity, and sometimes the direct experience of insanity, a spur to eloquence—*The*

art, literature, and music had long been preoccupied with representations of the suffering of the insane, and as the concert at Windsor
Castle showed, the King could be startlingly willing to enlist certain
of those representations on his own behalf.

Two of the passages selected by George III were easily applicable
to his situation: Samson's 'Total eclipse' (probably preceded by the
recitative 'O loss of sight'), and Jephthah's agonized lament 'Deeper
and deeper', as he realizes that he must kill his daughter to fulfil his
vow to God.[3] The King was, by 1811, almost completely blind, and
in November of the previous year his favourite daughter, the Princess Amelia, had died—a loss which was widely rumoured to have
plunged him irretrievably back into insanity.[4] Yet despite Horner's
claim that this was a programme devoted to madness and blindness, only one of the arias listed is, strictly speaking, a depiction of
insanity. No programme of the concert survives, but it is most likely
that the aria 'upon madness from love' was Dejanira's mad song
from the end of *Hercules*, sung when she discovers that her gift to
Hercules of Nessus' shirt, intended to reawaken his love for her, has
brought about his death.[5]

The applicability of 'madness from love' to George III's case is
not obvious. For many of the guests present, including the Duke of
Cambridge, it would in fact have been a more painfully apt choice
of subject than the King himself recognized. Princess Amelia ostensibly died of a wasting sickness (possibly tuberculosis), which produced acute nervousness in the final months of her life, but Princess
Mary voiced the belief of many closest to Amelia when she wrote to
the Prince Regent in January 1811 insisting that nervous susceptibility was not the symptom but the *'real cause'* of their sister's
death, 'for she died of a broken heart'.[6] Whether Princess Amelia's

---

Madhouse of Language: Writing and Reading Madness in the Eighteenth Century
(London, 1991), ch. 7 esp.; and on the problems of this emphasis on eloquence, see
my review in *History of Psychiatry*, 4 (1993), 456–8.

[3] In *Judges* 11: 31, Jephthah promises God that if he is glorious in battle he will
sacrifice the first living thing he meets. He meets his daughter.

[4] Dorothy Margaret Stuart, *The Daughters of George III* (London, 1939), 377.

[5] The repertoire for recitals of Handel had been so drastically reduced by 1811 as
to leave only a few likely candidates for the unnamed aria: 'Verdi prati' from *Alcine*,
'Sorge infausta' from *Orlando*, or Dejanira's aria from *Hercules*. The first two were
more popular choices for concerts of ancient music, but neither is strictly about
madness from love; Dejanira's aria was less often performed but more nearly fits
Horner's description. I am indebted to Richard Luckett for this information.

[6] Quoted in Stuart, *Daughters of George III*, 377.

love affair with the King's aide-de-camp Sir Charles FitzRoy was ever formalized in a clandestine marriage is not known, but it is unlikely that George III would have tolerated such a match. Though the Princess left everything to FitzRoy in her will, bar a few mementoes for friends and relatives, her brothers (including the Duke of Cambridge) acted as her executors and persuaded FitzRoy to forgo his claim to the property.[7] It is assumed that George III therefore never discovered the seriousness of the liaison and that he chose to hear an aria 'upon madness from love' without knowing how closely it impinged upon his family.

There was, as Francis Horner noted, something 'singular' and something 'very affecting' about the King's soliciting the public sympathy at all. Antony Storer had remarked on the singularity of George III's madness much earlier, writing to William Eden, Ambassador Extraordinary in Madrid, on 24 February 1789, after the King's recovery from his first illness, that 'If his Majesty's disorder was singular and surprising, his cure is not less so.'[8] The concert spoke to an unusually heavy burden of affliction upon a uniquely placed man, and George III's willingness to 'dwell upon the worst circumstances of his situation' suggests that his singularity and his pathos were connected in his own mind: in Jane Austen's lightly caustic phrase, it is, after all, 'singularity which often makes the worst part of our suffering'.[9] It is also his evident uniqueness which makes George III's significance in the cultural and institutional history of English madness peculiarly difficult to assess. Though his illness undoubtedly raised the prominence of public debate over insanity, and did much to help the prestige of medicine for the insane, his is an uncomfortably hyperbolic presence in the history of insanity. The mad king overdetermines the subject of madness in this period, prompting one historian to quip that Clio has been 'cunning'.[10] Opera if anything magnified the degree to which George III's case was outside the bounds of ordinary experience, casting him not with any 'type' of madness, but along-

[7] Stuart, *Daughters of George III*, 380–4.

[8] Quoted in Ida Macalpine and Richard Hunter, *George III and the Mad-Business* (London, 1969), 86.

[9] *Persuasion* (1817; Oxford, 1990), 18.

[10] Roy Porter, *Mind Forg'd Manacles: A History of Madness in England from the Restoration to the Regency* (London, 1987), 14–15: 'The cunning of history had its say too, driving King George III mad.' Also, p. 32 on the 'fortuitousness' of the King's madness for the advancement of public medical inquiry into insanity.

side tragic individuals *in extremis*: Samson, Jephthah, possibly Dejanira.

There was, in the proper sense of hyperbole, something 'over-reaching' about the King's engagement with his own madness on the Duke of Cambridge's night. Through the concert, George III made his insanity the occasion for an artistic display at which he, along with his court, was one of the audience. Opera raised the subject of the King's derangement but kept it still at a safe distance. It solicited pity, but did so discreetly, alluding only tangentially to insanity. 'God save the King' was the only direct statement here and it alone seems to have said too much, driving oblique connections suddenly uncomfortably close: 'mak[ing] sure the application of all that went before'. George III's public acknowledgement of his debility happened in excess of its actual subject, in heightened forms, in magnified diction, and he evidently found consolation in this hyperbolic displacement of his suffering into the register of the stage. More generally, music had a therapeutic effect on the King which his doctors saw and approved, encouraging him to play the flute and the harpsichord in his calmer moments (he had Handel's harpsichord at Windsor).[11] The concert therefore both solicited sympathy and was in itself a potential balm or tonic; but here the gap between medical representations of the King's condition and contemporary artistic representations of insanity becomes apparent. The non-mimetic function of representation, in this instance, is essential to its therapeutic value. If music heals it is because it offers the King a transformed vision of insanity, one which makes of madness something heroic, passionate, even sublime.

Hyperbole's links with insanity go at least as far back as classical roman rhetoric.[12] When George Puttenham came to illustrate

---

[11] Macalpine and Hunter, *George III and the Mad-Business*, 89, 151, 170. Robert Jones, not one of the King's own medical attendants, but author of the first medical commentary on the King's breakdown in 1788, had also recommended music as an aid to the recovery of mental and bodily vigour ('provided it is not too loud'). Jones, *An Enquiry into the Nature, Causes and Termination of Nervous Fevers; Together with Observations Tending to Illustrate the Method of Restoring His Majesty to Health and of Preventing Relapses of his Disease* (London, 1789), quoted in Macalpine and Hunter, *George III and the Mad-Business*, 101.

[12] Aristotle notably does not link hyperbole with madness, though its connection with the broader concept of the irrational is often implicit in *On Rhetoric* and the *Poetics*. He does, however, link hyperbole with anger and both with immaturity. See Patricia Parker, *Literary Fat Ladies: Rhetoric Gender, Property* (London, 1987), 38.

hyperbole in *The Arte of English Poesie* (1589), he noted its origins in the Greek *huperbolé*, the 'overreacher', but also its equivalent in the Latin '*Dementiens* or the lying figure'.[13] Puttenham also demonstrated, though he did not theorize, a gendered axis to hyperbole. He chose three contexts in which to illustrate the proper use of this difficult rhetorical ornament: praise of one's king, praise of one's mistress, and, in marked contrast, female complaint.

. . . as a certaine noble Gentlewoman lameting at the unkindnesse of her lover said very pretily in this figure.

> But since it will no better be,
> My teares shall never blin:
> To moist the earth in such degree,
> That I may drowne therein:
> That by my death all men may say,
> Lo weemen are as true as they.[14]

When hyperbole moves from men to women (from the king to the gentlewoman) in Puttenham's account, it moves also from praise to complaint, and on the woman's tongue it retrieves the traces of its Latin link with dementia, becoming the natural figure for a grief suicidal in its extremity. This is a despair which the speaker claims will literally overwhelm her, drowning her in her own tears. It is a 'pretty' use of the figure, but it is also complaint with a purpose: complaint against one man, directed at *all men* with a degree of determination, perhaps even with the hint of a threat. By the excesses of her passion, this woman will demonstrate to men that 'weemen are as true as they'. The verse is, of course, critically at odds with itself here, for if women are 'as true' as men in this speaker's experience, they must be false. 'Overreaching' the mark, hyperbole reveals itself in the woman's voice as, indeed, a 'lying figure', a distorted truth; or, more accurately, it exposes a gendered inflexion to the rhetoric of 'truth'. Feminine truth is of necessity hyperbolic for the gentlewoman, able to prove her grief only by what Puttenham called 'immoderate excess': to be 'as true' as men, she must surpass them not only in her fidelity (which goes without saying) but in the expression of it.

---

[13] George Puttenham, *The Arte of English Poesie*, ed. Gladys Doidge Willcock and Alice Walker (Cambridge, 1936; repr. 1970), 191.

[14] Puttenham, *Arte of English Poesie*, 193.

The authenticity of female complaint is notoriously suspect. The woman's lament is typically overheard and ventriloquized by a male voyeur. It produces intricate, arguably inextricable interdependencies between male and female voices, between a femininity predicated on passive suffering, and a masculinity which defines itself only tacitly in contradistinction to this damaged version of the feminine. Perhaps most fundamentally, it operates in the treacherous gap between testimony and description.[15] In yoking hyperbole to female complaint, *The Arte of English Poesie* has a direct descendant in one of the most influential Georgian works on rhetoric, Hugh Blair's *Lectures on Rhetoric and Belles Lettres* (1783).[16] Blair was among the most vigorous proponents of universal grammar and brought to his analysis of hyperbole a rubric for linguistic propriety which made questions of authenticity of voice, if not more pressing, certainly more didactic.[17] Blair, like Puttenham, turned to the image of a deserted, grief-distracted woman to illustrate hyperbole, and to support an argument for its proper employment:

When a Poet is describing an earthquake or a storm, or when he has brought us into the midst of a battle, we can bear strong Hyperboles without displeasure. But when he is describing only a woman in grief, it is impossible not to be disgusted with such wild exaggeration as the following, in one of our dramatic Poets:

> ——————— I found her on the floor
> In all the storm of grief, yet beautiful;
> Pouring forth tears at such a lavish rate,
> That were the world on fire, they might have drown'd
> The wrath of Heaven, and quench'd the mighty ruin.   LEE

This is mere bombast. The person herself who was under the distracting agitations of grief, might be permitted to hyperbolize strongly; but the spectator describing her, cannot be allowed an equal liberty: for this plain reason, that one is supposed to utter the sentiments of passion, the other speaks only the language of description, which is always, according to the

---

[15] John Kerrigan (ed.), *Motives of Woe: Shakespeare and 'Female Complaint'. A Critical Anthology* (Oxford, 1991), esp. 11–13, 15–17.

[16] Renewed interest in Puttenham was stimulated by the general revival of rhetoric in the second half of the 18th cent., and in 1811 *The Arte of English Poesie* was reprinted as the first volume in Hazlewood's series of 'Ancient Critical Essays upon English Poets and Poësy'.

[17] Olivia Smith, *The Politics of Language 1719–1819* (Oxford, 1984), 21.

dictates of nature, on a lower tone: a distinction, which however obvious, has not been attended to by many writers.[18]

Hyperboles, for Blair, could be divided into two kinds and judged accordingly. They could be of the kind that are employed in description, as of an earthquake or storm, or they could be 'suggested by the warmth of passion'. 'The best by far', he was decided, 'are those which are the effect of passion,' because they are justified by it. Hence Blair's disdain for Nathaniel Lee's lines on a weeping woman: had she spoken the words herself, they would have been rendered natural by her strength of feeling, but spoken by the male spectator they become absurd, even distasteful. Blair perhaps did not recall that Nathaniel Lee had spent five years of his life, between 1684 and 1689, in Bedlam, had been readmitted in 1690, and died insane after escaping his keepers. It is likely that the recollection would not, in any case, have altered the rhetorical point. In Blair's conception of rhetoric, as in Puttenham's, it was female passion which 'naturalized' hyperbolic expression, making it the rhetoric not only of opera and oratorio, but of less heroic genres, including the novel.

George III may seem an odd starting-point for a discussion of female insanity in the novel, but his 'singular' choice to broach the subject of his own illness through an operatic aria 'upon madness from love' is on a par with a kind of gendered hyperbolic displacement which makes itself felt continually in early nineteenth-century English fictional representations of madness. In an age which faced repeated constitutional crises in the King's derangement, there were few if any conventional types of male insanity which exercised a hold on the public imagination comparable with that of women's love-madness. The melancholic scholar and the raving bestial maniac still made appearances, but their heyday had been the mid-eighteenth century[19] and they proved far less attractive to the sentimentalists and their successors than they had to the Augustan

[18] Hugh Blair, *Lectures on Rhetoric and Belles Lettres*, 2 vols. (London, 1783), i. 320–1. Blair was a minister of the High Church of Scotland, and Professor of Rhetoric and Belles Lettres at Edinburgh University, and his *Lectures* enjoyed considerable success well into the 19th cent., going through 14 editions between 1783 and 1825, and several more by the 1860s. The quotation is from Nathaniel Lee's *Mithridates King of Pontus, A Tragedy* (1678), I. i. 153–9 (Blair omits two lines).

[19] See Michael V. DePorte, *Nightmares and Hobbyhorses: Swift, Sterne, and Augustan Ideas of Madness* (San Marino, Calif., 1974).

satirists. With no sympathetic terms in which to depict the subject of male insanity, Georgian and early Victorian writers increasingly viewed it through the harsh lens of grotesque comedy: a mode epitomized in Gillray's cruel caricatures of the King, and carrying through in fiction to such popular works as Samuel Warren's *Ten Thousand a Year* (1841) and Robert Surtees' *Handley Cross* (1843). Very rarely a writer did attempt a more subjective account of male madness—most notably James Hogg in *The Private Memoirs and Confessions of a Justified Sinner* (1824) and the late short story 'Strange Story of a Lunatic' (1830)—and the result put realism in complete disarray.

By comparison, the lovelorn madwoman was a far easier subject to contemplate, yet she too raised real anxieties for a culture increasingly suspicious of a sensibility which it had begun to identify with sentimental or revolutionary idealism.[20] The resulting tensions are acutely in evidence in early nineteenth-century fiction. In many respects, femininity and madness were so closely intertwined in English thinking as to be not easily separated. The love-mad woman also brought into peculiarly sharp focus the implications of rejecting wholesale a humane interest in the desolation of betrayal: those cadences of complaint to which eighteenth-century sentimentalism had been particularly atuned. 'Only a woman in grief', the love-mad woman enabled readers to contemplate madness as a subject of pathos rather than of terror, and as such she had much to offer a society which prided itself (not always justly) on its humanitarian advances; yet she was also unmistakably reminiscent of the worst excesses of sentimentalism. The resulting doubts about the legitimacy of the love-madness convention and the desirability of giving it further credence were played out time and again in fiction of the period, most obviously—and perhaps most pressingly—for women writers. What could women's relationship be to a convention now so distrusted, but which had at least admitted women's 'truth', the authenticity of their passion and their complaint, albeit only in a distorted aspect, only under the sign of hyperbole?

In the same month of 1811 in which George III planned his concert of excerpts from Handel, a new novel about women and madness

[20] See Marilyn Butler, *Jane Austen and the War of Ideas* (1975; rev. edn. Oxford, 1987), esp. 7–56; Warren Roberts, *Jane Austen and the French Revolution* (London, 1979).

was being advertised in the bookshops, and another was in the press. *The Passions*, a four-volume epistolary novel by the glamorous and prolific romantic novelist 'Rosa Matilda', appeared in May while Jane Austen was impatiently awaiting the appearance of her first published novel, *Sense and Sensibility*, due out in October.[21] It would be difficult to find two more polarized approaches to the sentimental vogue for love-mad women. *The Passions* exploits the gothic and sentimental taste for fiction about female insanity whole-heartedly; *Sense and Sensibility* is as determinedly an exposé of all that is deplorable in that taste. Yet even *The Passions*, which appears entirely unreconstructed in its fascination with women who lose their minds when they lose their men, has something to say about the implications and the limitations of its own conventionality. Equally, reading Austen alongside it underlines the attraction which the love-madness narrative continued to exert even for a writer deeply sceptical of its accompanying sentimentalism.

'Rosa Matilda' was the pseudonym of Charlotte Dacre, memorialized none too flatteringly by Byron in his 'English Bards and Scotch Reviewers' (1809):

> Far be't from me unkindly to upbraid
> The lovely ROSA's prose in masquerade,
> Whose strains, the faithful echoes of her mind,
> Leave wondering comprehension far behind.
> Though Crusca's bards no more our journals fill,
> Some stragglers skirmish round the columns still,
> Last of the howling host which one was Bell's,
> Matilda snivels yet . . .   (lines 755–62)[22]

Byron identifies Charlotte Dacre as one of the Della Cruscan school on the strength of her 1805 volume of poetry *Hours of Solitude*.

---

[21] See her letter to Cassandra Austen, 25 Apr. 1811, in *Jane Austen: Selected Letters 1796–1817*, ed. R. W. Chapman (London, 1955; Oxford, 1985), 113–18: 114.

[22] *Lord Byron: The Complete Poetical Works*, 7 vols., ed. Jerome J. McGann (Oxford, 1980–93), i. 253. Bell is John Bell (1745–1831), the English publisher of Robert Merry and the Della Cruscans. Ironically, Dacre's poem 'The Mountain Violet' was on several occasions misattributed to Byron, presumably on the mistaken assumption that it was a parody. See Henry T. Wake, 'Byroniana', *Notes and Queries*, 8th ser., 6, 25 Aug. 1894, 144–5. Byron parodied the *Hours of Solitude* under the title *Hours of Idleness*.

The Della Cruscans were a fervently Romantic group of poets, led by Robert Merry and given patronage by Mrs Piozzi. They made their name with a volume of poems written in Florence and published as the *Florence Miscellany* in 1783. Despite initial popularity, the tide of critical taste had turned severely against them by 1809.[23] 'Rosa Matilda' was, to moderate Byron slightly, one of the last of the howling host of gothic sentimentalists, and *The Passions* was her swan-song. The details of Charlotte Dacre's life are far from clear, but she was probably 39 in 1811, and she had already published three novels in addition to the early collection of poetry.[24] *Confessions of the Nun of St Omer* (1805) was dedicated to 'Monk' Lewis and, like *Zofloya; or, The Moor* (1806), *The Libertine* (1807), and *The Passions*, closely resembles Lewis's fiction in its extremity of rhetoric, plot, and passion. Unlike Lewis's work, however, the novels are remarkable for their strong women.[25] Each features at least one woman who becomes dangerous in pursuit of her sexual desires and who determinedly seeks revenge against those who stand in her way. In this respect Dacre was, as Ann H. Jones has argued, far from conventional.[26]

Critically, *The Passions* was a failure, making Dacre one of the more spectacular casualties of the new conservatism. The *Critical Review* found it 'deformed' by Dacre's 'inflated extravagance of diction' and 'impaired by the repulsive affectation of her style'[27]— a judgement whose harshness was in line with Dacre's usual reception, not only by Byron but by the majority of periodical reviewers. Kinder notices mixed flattering reference to her appearance with their remorseless invective on her work.[28] *The Passions* was also Dacre's least successful work in publishing terms. Sandra Knight-

[23] See W. N. Hargreaves-Mawdsley, *The English Della Cruscans* (The Hague, 1967); and E. E. Bostetter, 'The Original Della Cruscans and the Florence Miscellany', *Huntington Library Quarterly*, 19 (1956), 277–300. Jerome McGann's *Oxford Anthology of Romantic Period Verse* (Oxford, 1993) gives considerable space to the Della Cruscans.

[24] Ann H. Jones, *Ideas and Innovations: Best Sellers of Jane Austen's Age* (New York, 1986), 226–7; and Montague Summers, *Essays in Petto* (London, 1928), 57–8.

[25] See Robert Miles, *Gothic Writings 1750–1820: A Genealogy* (London, 1993), 179–88, on *Zofloya*'s interest as a 'female version' of *The Monk*.

[26] Jones, *Ideas and Innovations*, 228. Jones is writing specifically of the heroine of *Confessions of the Nun of St Omer*.

[27] The appendix to the *Critical Review*, 3rd ser., 24, 1 Dec. 1811, 550.

[28] *Hours of Solitude* (London, 1805) had contained a very attractive engraving of the author by Buck.

Roth claims in her foreword to the Arno Press reprint of *The Passions* that Dacre's works 'circulated briskly from the shelves of the circulating libraries',[29] but this novel appears in only a few of the circulating-library catalogues in which it might have featured: Godwin's at Bath, Horne's at Cheapside, and Brown's in Aberdeen (where it was still being advertised in 1821).[30] Unsurprisingly, it is absent from the more 'professional', male-sponsored subscription libraries of the period. Nor does there appear to have been a further edition of the novel, though all Dacre's earlier works had been reissued soon after their first publication and two had been translated into French.[31] The absence of further editions of *The Passions* probably reflected a change in the confidence of her publishers, as much as a change in the taste of her readership. Cadell & Davies was a well-established firm, popular but respectable, fashionable but polite: as their historian puts it, 'steady and unsensational'.[32] Dacre's last novel probably strayed too far from that middle line.

Dacre's earlier novels all depict women ravaged by love-madness, but her last work is unrelievedly obsessed with the subject. The plot of *The Passions* is directed by the Countess Appollonia Zulmer, 'monster of artifice' and 'shameless disappointed wanton', who finds herself rejected by the object of her desires, Count Wiemar. In revenge she sets about ruining his marriage to the beautiful and innocent Julia. Appollonia succeeds in entangling Wiemar's wife emotionally, though not in fact sexually, with another man, but when pressured to elope with him, Julia panics, confesses to her husband, and flees to Switzerland. She is eventually found by her husband's friend and confidant, Baron Rosendorf, but remorse and grief have driven her incurably insane. Julia is brought back to a cottage near Wiemar's home where she slowly declines. Every day she is secretly watched from the woods by her tormented husband,

[29] Sandra Knight-Roth, Foreword to Arno Press edition of *The Passions* (New York, 1974), p. vii. See also her 'Charlotte Dacre and the Gothic Tradition', doctoral dissertation, Dalhousie University, Nova Scotia, 1972.

[30] *Catalogue of the Public Library, Broad-Street, Aberdeen* (1821), 275 (Item 5995). I am indebted to Peter Garside for the information relating to holdings of Dacre's novel by circulating libraries. See his forthcoming bibliography of English novels 1780–1830, co-authored with James Raven.

[31] *Zofloya* had had similar treatment from reviewers but enjoyed an 'extraordinary vogue'. Summers, *Essays in Petto*, 62.

[32] Theodore Besterman (ed.), *The Publishing Firm of Cadell & Davies: Select Correspondence and Accounts 1793–1836* (Oxford, 1938), p. xiv.

whose pride will not allow him to forgive her and take her back. On a bitterly cold night, the dying Julia escapes her companion-keeper and staggers barefoot through the snow to expire on her husband's doorstep, reason tragically returning just as she dies. There Wiemar finds her in the morning and the novel closes as the Count himself descends into the madness of grief and remorse.

Dacre's portrayal of the maddened Julia fits the established sentimental patterns for representing the madwoman. Baron Rosendorf's letter to Wiemar describes Julia as she is found in Switzerland:

Heavens! what a scene presented itself to my view—how shall I describe it to thee? The wretched hut divided into two compartments, exhibited in one a miserable bed, if bed it could be called, which was composed of straw and leaves, and covered with a tattered rug. Reclined on this—can I, dare I tell it thee, lay the form, still lovely in ruin, of the *Wife of Wiemar*! Her features were sunk and wan, yet her cheeks were tinged with a carnation glow; her eyes darted painful brilliancy as they wandered round, her hair hung partly loose, and partly was bound up with yellow and withered leaves. She held in her hand a withered branch of the forest which she was waving to and fro. Her wretched attire hung in tatters, yet fantastic and wild, still betrayed in its disposition the disorder of her mind.   (iv. 147–8)

The signs are immediately legible: disordered dress means a disordered mind. Julia is distinctly Ophelian, her appearance according fairly closely with contemporary theatre's staging of Ophelia as a sentimentalized love-melancholic: 'still lovely in ruin'. Mary Bolton played Ophelia in 1811 and like other actresses of the period did so in 'decorous style, relying on the familiar images of the white dress, loose hair, and wild flowers to convey a polite feminine distraction'.[33] Dacre's madwoman nevertheless retains elements of an Ophelia the theatre was taking care to expunge—a less decorous woman whose insanity is openly linked to a disturbance of the erotic passions. Like that other Ophelia, Julia combines sexual knowledge and innocence, the signs of her sickness mimicking those of prostitution: her eye wanders, her dress is torn and wild, her hair hangs loose and dishevelled. The symptoms of insanity are primarily cosmetic (fever gives her cheeks 'a carnation

---

[33] See Elaine Showalter, 'Representing Ophelia: Women, Madness, and the Responsibilities of Feminist Criticism', in Patricia Parker and Geoffrey Hartman (eds.), *Shakespeare and the Question of Theory* (London, 1985), 82.

glow', and her eyes a 'painful brilliancy') and the erotic insuf-
ficiency of her clothing is repeatedly noted: when the forester discovers her almost perished in the snow, the 'tattered garb' which
hangs around her 'scarcely suffice[s] to shield her delicate body
from exposure'; her dead body, recovered from the snow at the end
of the novel, is 'half-naked' (iv. 153, 339). Sexual provocativeness
is, however, to some extent offset by a continual reference to
Christian iconography. When found 'half-naked' by a forester,
Julia's thin arms are 'crossed' upon her bosom, her feet are 'bare
and torn, and the blood which had flowed from recent wounds was
congealed upon them'. In death, the 'crimson stream' staining her
'bare and wounded feet' is again stressed, and the 'arms of frozen
alabaster' are flung above her head (iv. 152, 339). Allusion to the
crucifixion excuses the display of the madwoman's body, dispersing
the sexual implications of her exposure into a call for pathos.
Indecorous she may be, but derangement removes the taint of guilt,
making her a legitimate object for the man's solicitous gaze and his
solicitous touch.

Dacre's description of the insane Julia is most obviously conventional in its presentation of her as the object of a man's vision. The
madwoman, as Elaine Showalter remarks in another context, is
'exposed—and opposed—to the scrutiny of the man';[34] his rationality recognizes her derangement and his rejection of her is the
cause of it. Her madness is precipitated by loss of him and, in this
case, is obliquely her form of atonement for transgressing against
him. Julia's mental distress, like that of the medieval and Renaissance speakers of female complaint, is entirely documented by
men. The forester, Rosendorf, and, most obsessively, her estranged
and jealous husband, all in turn play the voyeur, secretly watching
the woman from a darkened or obscured position of vantage.
Describing her image and disclosing it to each other in letters is
an act of male solidarity: a solidarity whose sexual implications
are underlined in the central 'mad scene' by Julia's being 'exhibited'
on a bed. Once again, pity and titillation rapidly become
indistinguishable.

Dacre's Julia may serve as a gratifyingly crude application of the
hyperbolic terms of the sentimental love-madness convention, but it

---

[34] The idea of atonement through loss of reason is one of the main preoccupations
of Dacre's novel; e.g. Rozendorf's plea to Wiemar in Letter XCI: 'Receive, oh! yes,
receive the distracted wanderer, not the unworthy criminal' (iv. 175).

also provokes a number of questions about the kind of fantasy being played out in its pages. Not least, it raises the question of Dacre's own relationship to its sexual voyeurism. Her recycling of convention may seem unthinking, but the rest of the novel is interestingly troubled by the validity of this moment in its plot. Although it sentimentalizes the madwoman with undeniable crudity in this, its most extended depiction of insanity, elsewhere *The Passions* repeatedly undermines the spectacle of love-madness which it here appears to endorse.

It is, above all, crucial to the development of the novel that Julia's madness fulfils another woman's desire for revenge against Count Wiemar. In the difference between the voyeuristic account of Julia's love-madness and Appollonia's declaration of her 'mad' desire for revenge, there is a neat demonstration of Hugh Blair's distinction between good and bad hyperbole, dubious description and legitimate passion; but *The Passions* also moves beyond Blair in its concern to explore the relationship between these two hyperbolic forms. The crudely excessive depiction of Julia, maddened by guilt, is the product of Appollonia's revenge; inauthentically 'descriptive' hyperbole, to use Blair's terms, is impelled by a more 'permissible' hyperbolic imperative, voiced in the first person.

Appollonia Zulmer's undeniably hyperbolic diction is driven by a furious determination to turn her suffering back on the man who prompted it:

Ah! is it not maddening to be despised? When the wild throbbings of love are repaid with cold scorn, what feelings then agitate the heart? Can you not tell? When transports of extacy meet with repulse, what is the flash of horror which darts through the brain? Do you not know? When the object we adore doats on another, then, oh! then, can you not describe? This living hell *has* been mine, it is now *yours*. Judge then from the greatness of my revenge what must have been my love!' (iv. 85–6)

Dacre's villainous Countess succeeds in her intention. She drives first Julia, then Wiemar, to madness, with the result that female insanity becomes the key to a revenge which has the man as its final and more legitimate object. Julia's infidelity, one might have thought, would be sufficient retribution on Wiemar and neat enough in its symmetry, but Appollonia will not be content until the Count too has felt 'the flash of horror which darts through the brain'. Julia's madness is important only in so far as her suffering and eventually her death exact Wiemar's mental collapse. Appollonia's revenge thus frames the other woman's insanity as without value—even without meaning—except in so far as it affects the men who watch it. Indeed, the most revealing feature of female insanity in *The Passions* is the highly mechanistic way in which it functions as a means to eliciting *male* insanity. Hyperbole has a very clearly gendered axis here, just as it did in George III's case, and in Puttenham's and Blair's rhetoric. The woman's madness expresses hyperbolically a condition which belongs, more accurately, to the man, and it does so with the imperative not of pathos but of vengeance.

What does this hyperbolic displacement, this subordination of female insanity to the man's experience, mean for Dacre's writing about women? *The Passions*, for all its crudity, is acutely aware of the ambivalent role madness plays in the rhetoric of women's love, and, even in the service of revenge, the convention of love-madness proves far from enabling. In the first volume of *The Passions*, Appollonia's former governess, Mme de Hautville, learns of her ex-pupil's desire to be revenged on Count Wiemar, and warns her against it: 'Did I not always describe love to you as the most insane, and at the same time the most dangerous of all passions for a woman? Did I not always tell you, that love was the destruction of woman's power and glory?' (i. 61). Describing love as 'the most insane, and *at the same time* the most dangerous of all passions', Mme de Hautville gives voice to a problem with the credibility of love-madness that is, more properly, the problem of having no sufficiently compelling language through which to speak of women's love with moderation. The term 'madness' has become acutely problematic in its moral authority here. It is a powerful rhetorical means of registering a woman's sexual desire as significant, but it also makes that desire pathological. That the term 'dangerous' has become a more authoritative word between these women than 'insane' suggests how little the hyperbole of insanity gives women in the way of self-expression except in the self-destructive form of revenge.

Revenge is not finally endorsed by *The Passions*. Appollonia's craving for retribution makes her an ultimately monstrous figure whose mad passion bears all the hallmarks of syphilitic mania. When the Count discovers her hand in the ruin of his marriage, he curses her in terms which belong (somewhat elaborately) to the iconography of venereal disease:[35] 'may the vengeance of Heaven fall on her head!—may her rank blood be poured out on the earth!—may the birds of the air tear her piecemeal, and her mouldered bones, reduced to ashes, be scattered like the dust!' (iv. 100). Appollonia is eventually murdered at the instigation of a former servant and his mistress, and her brutally injured body does indeed pour its 'rank blood' 'copiously' upon the earth (iv. 115-16).

Through its two divergent representations of female madness— Julia's sentimental malady and Appollonia's vengeful mania— Charlotte Dacre's novel testifies to a problem which had coloured a great deal of gothic, Romantic, and sentimental fiction over the previous decades: that of whether female sexual desire can be rendered significant other than through the hyperbolic distortion which makes desire a synonym for madness. Like much of the popular romance fiction of the period, *The Passions* is remarkable for the violence of event and of rhetoric it finds necessary in order to explore sexual relations between men and women. The word 'madness', or such close equivalents as 'insanity', 'derangement', 'raving', occurs on almost every page of the novel. To love deeply is to love 'madly'; to suffer any intensity of emotion—grief, remorse, anger, jealousy, and, equally, hope, pleasure, devotion—is to risk mental disintegration. Dacre's men and women do not, however, have equal purchase on that rhetoric. Though Appollonia's revenge is assured at the end, in spite of her death, we never see Wiemar mad. For its final pages, Dacre's novel abandons the epistolary form and employs in its place a more objective authorial narration. The change is marked by the heading 'Conclusion', but what follows is a major portion of the narrative, describing Wiemar's increasing distraction, the escape and death of his wife, and his final descent toward insanity. The intensely subjective epistolary mode used thus far is apparently no longer appropriate, as if the man's madness, unlike the woman's, is not a fitting or a possible topic for subjective consideration. Wiemar comes across in his own letters as nobly, rather than pathetically, distressed. He is absolutely in command of his own psychological state, able to render it as controlled and effective literary sentiment. His ravings are the literary outpourings of a fine romantic sensibility, and when

[35] On the iconography of syphilis, see Sander Gilman, *Disease and Representation: Images of Illness from Madness to AIDS* (Ithaca, NY, 1988), 248–57.

his distress ceases to be articulate it moves beyond the vision of this novel. To make him manic would be to rob him of nobility and his 'tones scarcely human' are as close as Dacre comes to doing that to him. Wiemar's insanity is always announced in the future tense: 'I shall go mad!' he declares repeatedly, and Dacre draws the curtain at the point where that statement threatens to translate into reality. The novel ends as, blind with grief, he lifts the dead body of his wife from the snow, 'in all the distraction of returning hopeless love and of approaching madness!' (iv. 339–40). Wiemar's 'brain whirls', 'he raves in tones scarcely human', he is plunged into 'all the distraction of returning hopeless love', yet madness is somehow 'approaching' rather than fully present. The closest description Dacre offers of a man's madness is never quite authenticated. The truth of his insanity is still held at a distance.

In the last paragraph of the novel the author discreetly retreats:

Here, like the painter of old, we must draw a veil over woe we have not power to delineate; and trust to have impressed on the minds of those who have contemplated the picture we have offered, the danger of listening to the delusive blandishments of sophistry; of yielding to the guilty violence of the Passions, or of swerving even in *thought* from the sacred line of virtue, and our duty.   (iv. 340)

This final warning is hardly consistent with the rest of *The Passions*. Dacre's novel has been an extended exercise in yielding to the guilty violence of the passions, and her closing sentiments did not fool the *Critical Review*:

The moral, which the author draws at the conclusion of her performance is, we allow, widely different [from its content]; such a moral is not deducible from the premises before us, and it is no uncommon affectation to add sentences of this nature to the end of a novel, that it may pretend to be in some way subservient to the purposes of instruction.[36]

Unconvincing though the gesture towards propriety is, Dacre's statement underlines her ambivalence towards the fantasies she pursues. On the one hand, she persists with the hyperbolic diction of sentimental gothic tradition, using the language and postures of madness in order to assert the strength of women's desires. On the other hand, female madness emerges as a mechanistic device through which the villainess extracts her true revenge: the man's

[36] *Critical Review*, 3rd ser., 24, 1 Sept. 1811, 51–7: 52.

mental destruction. Dacre's closing assurance that madness is something she and her readers should abjure in real life is perhaps her clearest acknowledgement of the self-displayingly conventional character of love-madness. Female insanity is useful for this woman writer precisely because it is so much the stuff of sentiment and so evidently manipulable. By the end of the novel, the term 'mad' has, unsurprisingly, become extremely unstable, inhabited by deeply contradictory values, simultaneously desired and shunned. It is a pressure Dacre presumably felt it could sustain.

Jane Austen and Charlotte Dacre might seem polar opposites, but there are similarities between *The Passions* and *Sense and Sensibility*, and they are significant. Austen's novel, like Dacre's, is centrally concerned with the sentimental tradition of the love-mad woman. Her first published work of fiction, like Dacre's last, suggests that female insanity can be a displaced representation of a distress which belongs as much to men. Rather more reluctantly, Austen shares Dacre's awareness of how far an extravagant sentimentalism may be enabling for women as the only culturally persuasive language in this society for putting private feeling before public responsibility. And, perhaps most importantly in terms of where the convention would go next, Austen is acutely aware of the convention's ability to mean more than it says—above all, its capacity to act as a vehicle for revenge.

Sense and Sensibility had several close precursors as a critique of sensibility. As Park Honan has shown, Austen's first novel participates in a lively burlesque literature aimed at debunking the excesses of sentimentalism.[37] Her quick delight in the comic subversion of the convention is also evident from the frequency with which she made fun of it in her juvenilia. The earliest surviving story in the juvenile notebooks, 'Frederic and Elfrida' (probably written around 1790 when Austen was 14), contains a sharp parody of Crazy Jane balladry. Frederic, Elfrida, and friend have been wandering in a grove for 'scarcely . . . above 9 hours' when they are surprised by 'a most delightfull voice' warbling the following song:

> That Damon was in love with me
> I once thought & believ'd

[37] Park Honan, *Jane Austen: Her Life* (London, 1987), 288–9.

But now that he is not I see,
I fear I was deceiv'd.[38]

The walkers soon discover two elegant young women—the beautiful Jezalinda and her sister Rebecca, whose forbidding squint, greasy tresses, and hunched back, the reader is assured, do not detract from her charms. The lament is entirely gratuitous. It proves to have no bearing on the circumstances of either girl, and when, at the end of 'Frederica and Elfrida', another young woman throws herself into the deep stream running through her aunt's pleasure grounds in Portland Place it is not because she has been betrayed but because she has unthinkingly promised to marry two complete strangers on the same day. *Love and Freindship*, a boisterous burlesque of sensibility written when Austen was 16, treats the convention with even less respect. Confronted with the sight of their two husbands, both mortally wounded, Sophia swoons and Laura runs mad repeatedly:

'Talk not to me of Phaetons (said I raving in a frantic, incoherent manner)—Give me a violin—. I'll play to him & sooth him in his melancholy Hours—Beware ye gentle Nymphs of Cupid's Thunderbolts, avoid the piercing Shafts of Jupiter—Look at that Grove of Firs—I see a Leg of Mutton—they told me Edward was not Dead; but they deceived me—they took him for a Cucumber—' Thus I continued wildly exclaiming on my Edward's death. Hours did I rave thus madly and should not then have left off, as I was not in the least fatigued, had not Sophia who was just recovered from her swoon, intreated me to consider that Night was now approaching and that the Damps began to fall.[39]

Laura lives to appreciate the folly of such indulgence in sensibility. Sophia, less fortunate, is carried off by 'a galloping Consumption', brought on by fainting on wet grass.

Of all Austen's mature works, *Sense and Sensibility* most clearly continues the juvenile stories' interest in burlesquing love-madness. In its earlier version, *Elinor and Marianne*, written between 1795 and 1798, *Sense and Sensibility* has been more specifically identified as a parody of one rather poor example of the sentimental love-madness vogue, the novel *Mary De-Clifford*, published in 1792 and written by a friend of the Austen family, Sir Samuel Egerton

---

[38] Jane Austen, *Volume the First*, ed. R. W. Chapman (London, 1984), 7.
[39] Jane Austen, *Volume the Second*, ed. B. C. Southam (Oxford, 1963), 48.

Brydges.[40] But although specific allusions to Brydges' novel may surface in *Sense and Sensibility*'s depiction of Marianne Dashwood, the juvenile stories make it clear that Austen's target was not limited to any one novel.

Elinor and Marianne Dashwood both spend most of *Sense and Sensibility* suffering from their attachment to men they cannot have. Marianne falls very publicly in love with George Willoughby, who seems to return her affections unreservedly. When he unexpectedly deserts her, Marianne makes her distress equally public— all the more so when she learns that he has engaged himself to another woman. She declines rapidly into a state of hysterical debility, in the tradition of the sentimental heroines she models herself on. At the point where she seems doomed to madness or death, however, she is returned to health and married to the respectable Colonel Brandon, whose patient attachment she once disdained. By contrast, her sister Elinor suffers through romantic disappointment in silence, always presenting a calm face to society, and is finally rewarded with marriage to the man she loves, Edward Ferrars, now free from an earlier entanglement.

The novel's moral schema appears punitive, so far as Marianne is concerned, not least linguistically. 'Reflection', we are told near the end, has given 'calmness to her judgment'. Hugh Blair would have approved of the change: 'More or less of [the] hyperbolical turn will prevail in language, according to the liveliness of imagination among the people who speak it. Hence young people deal always much in Hyperboles. . . . Greater experience, and more cultivated society, abate the warmth of imagination, and chasten the manner of expression.'[41] Nevertheless, the novel evidently recognizes that sensibility provides an effective mode of communicating private feeling in the absence of any more viable idiom. Austen's engagement with the love-madness convention forms part of a more extended examination of the culture of sensibility. Sensibility helps to uphold the material distinctions of wealth and power in the society she portrays. Essentially, it serves a conservative function, registering and maintaining the divisions of class, power, property, and propriety. The novel opens with an examination of that process through a description of John Dashwood's failure to provide for his

[40] See Frank W. Bradbrook, *Jane Austen and her Predecessors* (Cambridge, 1967), 124–36.
[41] Blair, *Lectures on Rhetoric*, 319–20.

mother-in-law and sisters as charged to do by their dying father. Initially intending to fulfil his promise that the family will be 'comfortable', Dashwood soon persuades himself that less than the £3000 he had in mind will do. His wife, however, soon persuades him that even £500 a piece would be 'a prodigious increase to their fortunes'—and £100 would be 'comfortable' (44, 45). Between them, Dashwood and his wife gradually whittle down the sum until they persuade themselves that no money at all need change hands at this stage. What makes this supremely selfish performance plausible is the continual appeal to 'feeling'. John and Fanny Dashwood invoke the language of sensibility as a kind of currency in which the Dashwood sisters will, of course, repay whatever generosity is shown to them. Should he reduce his own wealth by giving to them, they will make reparation in feeling. Their gratitude will be 'raised', much as one raises taxes (46). So powerful is that language, Dashwood can eventually convince himself that no real transfer of money is necessary, his generosity of intention being a quite sufficient substitute for the deed.

Nervous distress holds a very high value within such a society. When the news breaks of Edward Ferrars's engagement to the socially inferior Lucy Steele, Mrs Jennings describes the effect on Fanny Dashwood:

she thought to make a match between Edward and some Lord's daughter or other, I forget who. So you may think what a blow it was to all her vanity and pride. She fell into violent hysterics immediately, with such screams as reached your brother's ears, as he was sitting in his own dressing-room down stairs . . . So up he flew directly, and a terrible scene took place, for Lucy was come to them by that time, little dreaming what was going on. Poor soul! I pity *her*. And I must say, I think she was used very hardly; for your sister scolded like any fury, and soon drove her into a fainting fit. Nancy, she fell upon her knees, and cried bitterly; and your brother, he walked about the room, and said he did not know what to do. Mrs Dashwood declared they should not stay a minute longer in the house . . . *Then* she fell into hysterics again, and he was so frightened that he would send for Mr Donavan . . . (260–1)

Every person in this lightly farcical scene uses a display of overwrought sensibility to defend their social standing where they see it imperilled. Mrs Dashwood's hysterics are plainly interpreted for us by Mrs Jennings as the effects of wounded vanity and pride, Fanny's mode of expressing outrage at an engagement so contrary to her intention that Edward should marry well ('some Lord's daughter or other'). Nancy Steele and John Dashwood testify to their own solidarity with and fear of Fanny by quieter demonstrations of sensibility (Nancy cries 'bitterly', while John Dashwood paces and declares himself at a loss), but nothing less than a full fainting-fit can begin to express Lucy's awareness of her culpability in having contracted an engagement so insulting to her hostess and so much above her place. The hysteria, fainting-fits, and floods of tears show a socially accepted and understood exchange of signs at work; but the excess of sensibility in the scene is comic because it is so clearly theatrical. Lucy, particularly, makes her way through the novel precisely by flattering her superiors with her *sensibility* of their superiority. By that means she captures Edward's brother Robert and achieves the social position she wants. Her 'respectful humility, assiduous attentions, and endless flatteries', we are told in the last chapter, have 'reconciled Mrs Ferrars to [Robert's] choice, and re-established him completely in her favour' (364).[42]

As Mrs Jennings's story makes apparent, nervous illness is revealed in behaviour rather than in speech. It testifies to the inadequacy of language, becoming active at the points where words are too difficult or too crude to express certain desires or motives. Austen depicts a society in which speech is so curtailed and conduct so circumscribed that surfaces have assumed a heightened capacity for signification.[43] Colonel Brandon 'talked of flannel waistcoats,' says Marianne, 'and with me a flannel waistcoat is invariably connected with aches, cramps, rheumatisms, and every species of ailment that can afflict the old and the feeble' (71). Her prejudice is the comic extreme of an interpretative practice in which everyone in her society participates. Possession of a person's image becomes

[42] Cf. George Rousseau's suggestion that nervous illness (particularly melancholy) is largely put on for fashionable effect by socially mobile women (Elizabeth Bennet is one of his examples). George Rousseau and Roy Porter (eds.), *The Ferment of Knowledge: Studies in the Historiography of Eighteenth-Century Science* (Cambridge, 1980), 205.

[43] Austen scholarship has participated in the recent burgeoning of interest in the body. See Jan Fergus, 'Sex and Social Life in Jane Austen's Novels', in David Monaghan (ed.), *Jane Austen in a Social Context* (London, 1981); Toby A. Olshin, 'Jane Austen: A Romantic, Systematic, or Realistic Approach to Medicine?', *Studies in Eighteenth-Century Culture*, 10 (1981), 313–26; Claudia L. Johnson, 'A "Sweet Face as White as Death": Jane Austen and the Politics of Female Sexuality', *Novel*, 22 (1989), 159–74; Judy van Sickle Johnson, 'The Bodily Frame: Learning Romance in *Persuasion*', *Nineteenth Century Fiction*, 38 (1983), 43–61; and, most recently, John Wiltshire, *Jane Austen and the Body* (Cambridge, 1992).

tantamount to possession of the person. Lucy Steele can claim a real
advantage over Elinor in owning Edward's miniature. Whether or
not she has his affection does not matter to her, because she has the
sign of it. Possession of a lock of hair carries the same significance,
as does possession of a person's 'hand', or writing. So, her correspondence with Edward reinforces Lucy's claim to his hand in
marriage; a correspondence between Marianne and Willoughby
will be, for Elinor, the proof of their engagement; and the return of
Marianne's letters with the lock of her hair is the final proof that
Willoughby has betrayed her affection. When a stolen lock of hair
is a sufficient substitute for a violated body, no wonder that fetishism becomes the predominant mode for thinking and speaking of
sexual desire. Communication, for the Dashwood sisters, depends
upon their ability to read the expression of feeling in the slightest of
tokens.

In *Sense and Sensibility* the fascination with hysterical illness is,
more accurately, a fascination with the relationship between desire
and language. What kind of voice is available to women in a society
which instructs them to break down rather than declare intentions
of their own? Through Marianne Dashwood, *Sense and Sensibility*
investigates the possibility of using the signs and symptoms of
nervous sickness to register the double distress of the woman who
loves impossibly or inappropriately, and who must be silent under
that distress. Her nervous illness, like Lucy Steele's fainting-fits and
Fanny Dashwood's hysterics, is a mode of negotiating her desires in
a manner legible and acceptable to society, but Marianne differs
from those women in the lengths to which she will manipulate
sickness. With her, illness begins to express a more dangerous
degree of private interest and private protest.

Unexpectedly deserted by Willoughby, Marianne stages her distress in terms absorbed from the sentimental literature she reads
(we know, from the first chapter, that she is particularly affected by
Cowper—a more extreme expression of Austen's own taste):

[The family] saw nothing of Marianne till dinner time, when she entered
the room and took her place at the table without saying a word. Her eyes
were red and swollen; and it seemed as if her tears were even then restrained with difficulty. She avoided the looks of them all, could neither eat
nor speak, and after some time, on her mother's silently pressing her hand
with tender compassion, her small degree of fortitude was quite overcome,
she burst into tears and left the room.

This violent oppression of spirits continued the whole evening. She was without any power, because she was without any desire of command over herself. The slightest mention of any thing relative to Willoughby over-powered her in an instant; and though her family were most anxiously attentive to her comfort, it was impossible for them, if they spoke at all, to keep clear of every subject which her feelings connected with him.   (109)

Uncontrollable tears, lack of appetite, inability to speak, and a tendency to rush in distress from the room are all part of a display of mental anguish which becomes more dramatic as Willoughby proves to have betrayed her. The reader is clearly intended to check any welling of sympathy for this violent 'oppression of spirits'. The phrase itself is ambiguous, for Marianne is both oppressed and oppressing. Rather than being the victim of an uncontrollable sensibility, she is represented for the most part as responsible for her own mental state: she is powerless because she does not choose to assert power; her tears 'seem' rather than 'are' restrained with difficulty. 'Misery such as mine has no pride,' she tells Elinor, and her speech, littered with dashes and dominated by the first-person pronoun, is immediately reminiscent of sentimental fiction (with uncomfortable implications for the speaker): 'I care not who knows that I am wretched. The triumph of seeing me so may be open to all the world. Elinor, Elinor, they who suffer little may be proud and independent as they like—may resist insult, or return mortification—but I cannot. I must feel—I must be wretched—and they are welcome to enjoy the consciousness of it that can' (201). *Sense and Sensibility* continually reminds us that Marianne is *acting* the role of suffering romantic heroine. Always the same charge is laid against her, explicitly and implicitly: that she indulges, feeds, courts her grief, inflicts it on other people, indeed that she can measure it by the pain it gives to others. Far from powerless, she is, evidently, 'potent' and using her power selfishly.

Nevertheless, *Sense and Sensibility* remains confused in its criti-cism of her conduct. It seems often curiously divided in its attempts to find her culpable. Most obvious is the difficulty encountered in dealing with the status of Marianne's feeling. The description of her reaction to Willoughby's sudden departure ironizes and deprecates the selfish cultivation of her own suffering, yet the irony is con-fused, for the passage none the less makes a claim for the sincerity of her feelings:

Marianne would have thought herself very inexcusable had she been able to sleep at all the first night after parting from Willoughby. She would have been ashamed to look her family in the face the next morning, had she not risen from her bed in more need of repose than when she lay down in it. But the feelings which made such composure a disgrace, left her in no danger of incurring it. She was awake the whole night, and she wept the greatest part of it. She got up with an headache, was unable to talk, and unwilling to take any nourishment; giving pain every moment to her mother and sisters, and forbidding all attempt at consolation from either. Her sensibility was potent enough!   (110)

Marianne does spend a sleepless night, does arise with a headache because her 'feelings' leave her 'in no danger' of doing otherwise. Then again, what those feelings are, and how admirable, is uncertain. The sentence which appears to elucidate them in fact leaves them far from clear: '*the feelings which made such composure a disgrace*, left her in no danger of incurring it'. Austen's language vacillates between presenting Marianne as actively cultivating a false sensibility and presenting her as the victim of a sincerely suffering sensibility. It is a dilemma which appears as much Marianne's own as it is the text's. That she was 'unwilling' to take any nourishment suggests a culpably calculated pose, but it is offset by the earlier 'she was *unable* to talk'. Similarly, at the close of the previous chapter, she actively 'avoided' their looks but passively 'could' neither eat nor speak.

Marianne's physical collapse promises to be the pinnacle of her achievement as self-constituted sentimental heroine. It is also the key moment in the novel's critique of this mode of self-fashioning. Her sickness is the result of spectacular 'imprudence':

Two delightful twilight walks...not merely on the dry gravel of the shrubbery, but all over the grounds, and especially in the most distant parts of them, where there was something more of wildness than in the rest, where the trees were the oldest, and the grass was the longest and wettest, had—assisted by the still greater imprudence of sitting in her wet shoes and stockings—given Marianne a cold so violent, as, though for a day or two trifled with or denied, would force itself by increasing ailments, on the concern of every body, and the notice of herself.   (302)

The impetus is cumulative: Marianne constantly seeks the authority of extremism for her desire: 'most distant', 'more of wildness', 'oldest', 'longest', 'wettest', 'still greater', 'so violent', 'would *force* itself by *increasing* ailments'. The logic of the progression echoes

Marianne's flirtation with illness: it continually extends the process of becoming sick, but it staves off the moment of actual illness. Constant interruptions hold that moment at a distance, and when it arrives it has to be forced on Marianne's own notice; for this is the critical moment in Marianne's history when she ceases to be able to control her nervous condition, and becomes 'genuinely' threatened by it (that there should be a distinction between authentic and inauthentic feeling is, of course, the claim at the heart of *Sense and Sensibility*).

The comparisons between Austen's sick-bed scene and Dacre's mad scene are numerous, but most significant is Austen's avoidance of any external description of the woman's derangement. Refusing to make her an attractive object for the reader's contemplation, she avoids the eroticization of female irrationality common to the sentimental convention.[44] For the reader, pathos is therefore likely to be supplanted by anxiety about the woman whose illness threatens to remove her from the narrative altogether. Representations of deranged women in literature of the period typically describe them as pathetically inarticulate but make them in practice very articulate indeed, able and more than willing to give heart-rending testimony to their desertion. Marianne, by contrast, almost loses her voice. The one coherent thing she says among her inarticulate cries is to demand not her lover but her mother: 'Is mama coming? . . . she must not go round by London . . . I shall never see her, if she goes by London' (306).

One striking consequence of Austen's rejection of sentimental convention is that Marianne becomes a more genuinely distressing figure after her recovery than she was when her sickness conformed to the sentimental pattern. The last chapters of the novel magnify a degree of uncertainty that has always been quietly present in the novel over how far Marianne is, finally, to blame for her selfish indulgence of sensibility, for she continues to be prey to symptoms of nervous strain which disturb the novel to a greater degree than before because, devoid of their customary context, they lack an interpretative framework. The close of chapter 46 has Marianne,

---

[44] Eroticization of the body of the sick woman is also a recurrent feature of medical and scientific texts of the period. See Ludmilla Jordanova, *Sexual Visions: Images of Gender in Science and Medicine between the Eighteenth and Twentieth Centuries* (New York, 1989); and Evelyn Fox Keller, *Reflections on Gender and Science* (New Haven, 1984).

now knowing of Willoughby's love for her, retire alone upstairs, weeping, scarcely able to speak. The symptoms of nervous distress are the same as they always were, but this time they appear to be out of Marianne's control and no easy source of egoistic gratification to her. Her sister's response underlines the change: 'Elinor would not attempt to disturb a solitude so reasonable as what she now sought' (339). But the tearful retreat is not Marianne's last. Her sympathy with Elinor's misery seems at times as wanting in control as her own unhappiness once was. Indeed, she increasingly appears to have displaced her own 'solitary emotional fantasies' into public expressions of sympathy for Elinor. When news arrives that Edward Ferrars is married to Lucy Steele, Marianne breaks down completely, and has to be taken from the room before Elinor's distress can be considered: 'Marianne gave a violent start, fixed her eyes upon Elinor, saw her turning pale, and fell back in her chair in hysterics' (343). It is a characteristically subtle Austenian irony that this display, dubious though it is in its own terms, has the positive effect of affording Elinor time to compose her feelings amid the rush of attention to her sister. When Edward makes his appearance, some days later, Marianne 'retreat[s] as much as possible out of sight, to conceal her distress' (349). Even in her calmer moments, she finds in the language of repentance a remarkably close replica of her old romantic extravagances of diction: 'I wonder', she tells Elinor, 'that the very eagerness of my desire to live, to have time for atonement to my God, and to you all, did not kill me at once' (337).

Marianne's near absence from the last chapter is symptomatic of the difficulty of absorbing this excess of self-punishment into the more benign channels of domestic felicity. We are told that she restores Colonel Brandon's 'mind to animation, and his spirits to cheerfulness' (367) and that she finds her own happiness in forming his, but the process of her restoration to mental health is elided. The last occasions of her nervous breakdown remain problematic and disturbing, unsentimentalized but also unresolved. The novel simply declares them cured. Buried in that resolution, however, is another happy ending, one which most critics have downplayed or overlooked altogether. Colonel Brandon is an awkward presence in *Sense and Sensibility*, never quite proof against Marianne and Willoughby's mockery or the taint of flannel waistcoats. Many critics have been openly hostile to this rather stolid figure whose love seems at last to be imposed on Marianne by the collective will of her family and friends. Brandon, however, has his own history of disappointed love, surprisingly similar to Marianne's and closely intertwined with it.

When Brandon asks Elinor early on, with a faint smile, whether her sister does not consider 'second attachments' an impossibility, Elinor's interest in this 'silent man of five and thirty' increases:

She liked him—in spite of his gravity and reserve, she beheld in him an object of interest. His manners, though serious, were mild; and his reserve appeared rather the result of some oppression of spirits, than of any natural gloominess of temper. Sir John had dropt hints of past injuries and disappointments . . . she had reason to suspect that the misery of disappointed love had already been known by him. (81, 86)

Rumours that Colonel Brandon has a natural daughter, a Miss Williams, in London reach the Dashwood circle, but the true story only emerges when Brandon feels obliged to give Elinor his reasons for distrusting Willoughby. The result is very much in the eighteenth-century tradition of the man of feeling. Brandon tells a story in a voice broken with emotion, and punctuated by heavy sighs. At the crisis of his story he is compelled to stop: 'He could say no more, and rising hastily walked for a few minutes about the room. Elinor, affected by his relation, and still more by his distress, could not speak. He saw her concern, and coming to her, took her hand, pressed it, and kissed it with grateful respect. A few minutes more of silent exertion enabled him to proceed with composure' (216).

Years before, Brandon fell in love with his childhood playmate, a girl named Eliza who was orphaned in infancy and brought up under the guardianship of Brandon's father. As the younger son, Brandon was not permitted to marry his sweetheart, because her fortune would come into the Brandon family only if she married his older brother. A planned elopement was thwarted at the last minute, Brandon was banished and Eliza forcibly married to the brother: a licentious man whose inconstancy eventually drove his friendless wife into infidelity. When Brandon returned from his regiment in the East Indies he learnt that she had been divorced on grounds of adultery. 'It was *that*', Brandon tells Elinor, 'which threw this gloom,—even now the recollection of what I suffered'

(216). He found Eliza, eventually, in a debtors' house, where she was dying of consumption, her health and her beauty wrecked by prostitution and poverty (a touch strongly reminiscent of *The Man of Feeling*): 'So altered—so faded—worn down by acute suffering of every kind! hardly could I believe the melancholy and sickly figure before me, to be the remains of the lovely, blooming, healthful girl, on whom I had once doated' (216). She died, leaving in his care a 3-year-old child 'the offspring of her first guilty connection' (217). Through this child, Brandon's story overlaps with Willoughby's. Shortly before the Dashwoods met Willoughby, he had met the girl at Bath, where she was staying with a schoolfriend, seduced her, then abandoned her 'in a situation of the utmost distress, with no creditable home, no help, no friends, ignorant of his address! He had left her promising to return; he neither returned, nor wrote, nor relieved her' (218). It almost always goes unnoticed that in fine eighteenth-century tradition, Colonel Brandon duels with Willoughby to avenge the honour of Eliza's daughter (though both men, significantly, escape unhurt; 220).

The story of Eliza and her 'wretched and hopeless' daughter, who still loves Willoughby but who will all her life have 'a mind tormented by self-reproach', is explicitly invoked as a warning for Marianne. But equally important is what this story says about Colonel Brandon. He tells his story sparely, soberly, and with a clear eye to its moral, but its classically sentimental content still sits rather oddly within a novel otherwise highly critical of sentimentalism. This is the Colonel's claim to a 'compassionate respect' which even Marianne now readily accords him. In one sense, it establishes just as strongly as Dacre's depiction of Wiemar the extent to which male sensibility is only authorized when it is displaced on to women. In another, it suggests just as strongly that feminine sensibility has become so debased for Austen that the man has a better claim on the convention of love-madness than the woman.

Rejecting the period trope of the madwoman is not just a matter of declining an uncongenial construction of femininity in *Sense and Sensibility*. It also entails a complex questioning of how women's emotional life is to be represented in the wake of that rejection. Marilyn Butler has argued influentially that this novel remorselessly punishes a sensibility it allies with 'sentimental (or revolutionary) idealism'.[45] At the same time, there is an unrelenting pressure in Austen's novel to find a diction which can more honestly express the demands of feeling without resource to sentimentalism. Austen refuses sensibility not just because she associates it, in general terms, with revolutionary idealism, but because (with a rather different impulse) she accepts how far the particular trope of female madness has been reduced in its capacity to register the cadences of a distress which is no less real for having been described so many times before. The novel asks us to recognize a fundamental distinction between the blind adoption of too familiar, too sentimentalized literary expressions for feminine distress, and an alternative pressure to find a more convincing representation of the psychological stress that is part of the everyday unseen lives of her women.

Jane Austen's interest in expressing desire and its dangers without subscribing to the values of sentimentalism is centred on her depiction of Elinor Dashwood. Critics have traditionally seen Elinor as an apologist for rationalism, unquestioningly endorsed by the novel, but LeRoy Smith is right to note that rationalism is never materially empowering for her, and that she has to rely on a late twist in the plot for her self-sacrificing sense to be rewarded.[46] Marianne charges her sister at one point with refusing or failing to 'communicate' (184), but, while Elinor does not draw public attention to her love for Edward, her pain does not go unregistered. As Gillian Beer has remarked, Elinor's supposedly rational diction bears signs of strain. Her speech typically sets up complex constructions of reasoned argument which then fail to be accomplished. Her voice repeatedly embroils itself in confused webs of syntax, or tails off altogether.[47] At the same time, this language is remarkably unornamented. Eschewing hyperbole, Elinor's speech is almost as

---

[45] Butler, *Jane Austen and the War of Ideas*, 194, 192: 'It is the role of Marianne Dashwood, who begins with the wrong ideology, to learn the right one. . . . It is quite false to assume that merely because Marianne is treated with relative gentleness, Jane Austen has no more than a qualified belief in the evils of sensibility. She spares Marianne, the individual, in order to have her recant from sensibility, the system.' For a further, less convincing example of critical resistance to valuing the text's repressions above its ostensibly conservative conclusion, see LeRoy W. Smith, *Jane Austen and the Drama of Women* (Totowa, NJ, 1983), 80–6.

[46] Smith, *Jane Austen and the Drama of Women*, 85.

[47] See e.g. her extraordinarily contorted disclaimer when Marianne charges her with loving Edward, *Sense and Sensibility*, 53–5; or her conversation with Edward, p. 349. Gillian Beer remarks upon this aspect of the novel, in 'Beyond the Sensible, Lucid Jane'.

spare of decoration as the novel's description of her (one brief sentence tells us that, in Willoughby's eyes, 'Miss Dashwood had a delicate complexion, regular features, and a remarkably pretty figure'; 78).

In place of the conventional hyperboles of sentimental fiction, *Sense and Sensibility* thus engages in a continual search for a language in which to say something more telling about the cost of being politely feminine in this society, and (it should not be forgotten) about its possible rewards. Certainly, the end of *Sense and Sensibility* exhibits a strongly felt wish to change what fiction can say about women in love. Austen's account of Elinor and Marianne's marriages itself marries the voices of sense and sensibility in an intricate linguistic comedy. So, Austen gives us the reasoned romance of Elinor and Edward:

after waiting some time for [the completion of improvements to the house], after experiencing, as usual, a thousand disappointments and delays, from the unaccountable dilatoriness of the workmen, Elinor, as usual, broke through the first positive resolution of not marrying till every thing was ready, and the ceremony took place in Barton church early in the autumn. . . . Mrs Jennings's prophecies, though rather jumbled together, were chiefly fulfilled; for she was able to visit Edward and his wife in their Parsonage by Michaelmas, and she found in Elinor and her husband, as she really believed, one of the happiest couple in the world.   (363)

The novel, like Mrs Ferrars, appears to be inspecting a happiness it is 'almost ashamed of having authorised' and, like her, retreats into comic oscillation between emotional engagement and a rationalizing detachment whose impulse is nevertheless transparently romantic. Austen teases her readers as she fulfils their expectations—and not the least part of the comedy is her insistence that her heroine's happiness is not exceptional (on the contrary, the repeated 'as usual' insists upon its ordinariness). When the novel offers the opinion that Elinor and Edward were, 'one of the happiest couple in the world', it puts the phrase in Mrs Jennings's mouth and, coming from her, the statement has the full effect of its double edge. It is both cliché and something more—affirmation: 'she really believed'.

The description of Marianne's marriage to Colonel Brandon is the more difficult passage to read. Plainly and avowedly a marriage of sense, Austen's gestures towards the conventions of romance fiction have seemed to many readers out of place when Marianne's emotions appear so coerced: 'With such a confederacy against her—with a knowledge so intimate of his goodness—with a conviction of his fond attachment to herself, which at last, though long after it was observable to everybody else—burst on her—what could she do?' (366). The terms of sense—knowledge, conviction, observation—vie with the terms of sensibility—intimacy, fond attachment, conviction 'bursting on her'. The dashes which punctuate the text, at once joining opposing tones and marking their opposition, reveal Austen's appropriation of one of the trademarks of sentimentalism, even as she denies the sentimental reader the pleasure of indulging in convention. The last phrase encapsulates the strain Austen is placing upon the language of rationalism, for it is tonally at odds with the lengthy process of reasoning that has gone before. A spontaneous overflow of sensibility invades the measured tones of sense, and the conclusion finally takes refuge from its own convolutions in a rhetorical question which is an appeal to both qualities: 'What could she do?' That question would go on making itself heard in Austen's writing, and it sums up her continuing desire to find a new language in which to explore the psychology of women betrayed in love.

'Rosa Matilda' and Jane Austen come together in their demonstration of the extent to which the story of a woman's 'madness from love' had become not merely conventional but unavoidably so. The attack on sentimentalism brought with it a new scepticism about the desirability of indulging readers with stories about women who lose their minds when they lose their lovers—a scepticism for which Austen is well known, but which shows even in the writing of Charlotte Dacre. At the same time, the convention of the love-mad woman was so much part of English culture, so powerful in its appeal even when sentimentalism had come into disrepute, that a curious tension emerges in early nineteenth-century fiction: a tension between the desire to reject the tradition altogether and a recognition of its usefulness *as* convention. It is under these conditions that stories about madwomen begin to take on a new meaning in the British novel: or rather, a new series of meanings. The peculiar mixture of pleasurable familiarity and critical distance which nineteenth-century writers and readers brought to this story made it perhaps uniquely well placed to serve as a vehicle for other less easily articulable concerns. The question both Charlotte Dacre and Jane Austen faced of the consequences of this convention for women becomes more pressing and at the same time more elusive in a climate where it was becoming easier to see that the love-mad woman's story was not necessarily about her at all, but about other voices and other narratives finding expression, hyperbolically, through her.

4

Love-Mad Women and Political Insurrection in Regency Fiction

IN 1828 George Man Burrows summed up the medical consensus of the previous four decades regarding the detrimental effects of insurrection on a nation's mental health:

Insanity bears always a striking relation to public events. Great political or civil revolutions in states are always productive of great enthusiasm in the people, and correspondent vicissitudes in their moral condition; and as all extremes in society are exciting causes, it will occur, that in proportion as the feelings are acted upon, so will insanity be more or less frequent.

Accordingly, Pinel has observed, how common mental alienation was in France, from the effects of the revolution, and Dr. Hallaran remarked the same, as the effect of the last rebellion in Ireland. Rush has given many examples of the influence of the American revolution on the human body and mind . . . These affections were so frequent among the royalists, that Rush gave them the specific name of 'Revolutiana,' and they bore the character of despondency; to the species of insanity pervading the revolutionists, that of 'Anarchia,' bearing the opposite character.[1]

As Burrows indicates, early nineteenth-century medical interest in the concept of revolutionary insanity was largely inspired by Pinel's work with the insane after the French Revolution; but concern about the phenomenon was still strong almost forty years later. Burrows employs a language of 'exciting causes' loosely indebted to a humoral pathology supposedly superseded by the nineteenth century but still traceable in its language:[2] a language of pathological 'influences', emphasizing the organism's susceptibility to suggestion. Revolution does not occur spontaneously within a human body, for Burrows, nor, by extension, does it occur spontaneously within the body politic: it is visited upon the system, 'exciting' it

---

[1] George Mann Burrows, *Commentaries on the Causes, Forms, Symptoms, and Treatment, Moral and Medical, of Insanity* (London, 1828), 20.

[2] Janet Oppenheim, *'Shattered Nerves': Doctors, Patients, and Depression in Victorian England* (New York, 1991), 88–9.

from without, inducing despondency in the conservatives and mania in the radical sympathizers. Clearly, this is the doctor in the role of public moralist, rather than that of practical physician. Indeed, the concept of 'revolutionary insanity' gains in impact from the sheer difficulty of imagining its translation into the treatment of individuals. 'Anarchia' and 'Revolutiana' have no cure.

Medical opinion about the dangers of political rebellion found ready support in anti-Jacobin writing of the 1790s. Edmund Burke's preferred metaphor for the French Revolution and its after-effects was that of sickness—a plague or palsy threatening England from across the Channel[3]—but the language of madness also appears repeatedly in the *Reflections*: 'Woe to that country which would madly and impiously reject the service of the talents and virtues . . . that are given to grace and serve it'; 'Governments must be abused or deranged indeed, before [the dethroning of kings] can be thought of.' In one of the more philosophical passages of the *Reflections* it becomes clear that Burke saw madness less as the cause of revolution than as the consequence of breaking the social contract. Once the path of revolution is chosen by a people, he wrote, society dissolves into chaos: 'the law is broken, nature is disobeyed, and the rebellious are outlawed, cast forth, and exiled, from this world of reason, and order, and peace, and virtue, and fruitful penitence, into the antagonist world of madness, discord, vice, confusion, and unavailing sorrow'.[4]

In pitting law against chaos, virtue against vice, reason against madness, Burke was attempting to set the rhetorical stakes for thinking about revolution in England. To a considerable degree he succeeded. John Turner has noted how, paradoxically, Thomas Paine was obliged to combat Burke's extreme invective against revolution by making his own defence of the Revolution contrastingly sober and restrained in its expression.[5] But though the association of madness with revolution tends now to be remembered as

---

[3] Edmund Burke, *Reflections on the Revolution in France and on the Proceedings in Certain Societies in London Relative to that Event* (1790; Oxford, 1993), 89.

[4] Ibid. 139, 117, 195.

[5] John Turner, 'Burke, Paine, and the Nature of Language', in J. R. Watson (ed.), *The French Revolution in English Literature and Art: The Yearbook of English Studies*, 19 (London, 1989), esp. 48–52. Tom Furniss develops this argument in 'Rhetoric in Revolution: The Role of Language in Paine's Critique of Burke', in Keith Hanley and Raman Selden (eds.), *Revolution and English Romanticism: Politics and Rhetoric* (Hemel Hempstead, 1990).

a characteristic of anti-Jacobin writing,[6] it had a much wider currency.

For medical writers, as Burrows indicates, revolutionary insanity could comprehend both depression and mania—manifestations of insanity usually considered typologically distinct in medical theory of the period—and its precise functioning was open to representation in quite different ways. In emphasizing a model of revolutionary insanity as contagion, for example, Burrows differs noticeably from his most recent historical source, William Saunders Hallaran's *An Enquiry into the Causes Producing the Extraordinary Addition to the Number of Insane* (1810). The emphasis in Hallaran's writing is upon madness springing up *within* Ireland, bred from its own political troubles. To account for the escalating incidence of madness, he claims, it is necessary to acknowledge 'the high degree of corporeal as well as mental excitement' resulting from 'continued warfare in the general sense' and, more particularly, from that state of warfare which '*engenders and brings to issue the horrors of intestine feuds*; imparting visionary views to some, "who build their hopes upon their country's ruin!" to others, all the pangs which follow quick upon licentious arrogance!—The one . . . at length gives way;—yet reigns pre-eminent on a Throne of Straw! The other . . . sinks into hopeless apathy.'[7] The first edition of Hallaran's work briefly footnotes the French Revolution as evidence of the link between political disturbance and the spread of insanity,[8] but the reference is entirely omitted in the revised edition.[9] History, in other words, has little of the portentous weight it possesses in Burrows's text, where historical precedent itself appears capable of infecting the present. Hallaran might have been expected to make rather more of the French example than he did. From its beginnings, militant Irish republicanism was closely bound to the example and the practical aid of the French;[10] but where,

[6] See John Brewer, ' "This Monstrous Tragi-comic Scene": British Reactions to the French Revolution', in David Bindman (ed.), *The Shadow of the Guillotine: Britain and the French Revolution (1789–1820)* (New Haven, 1983), esp. 38–40, 57–73, 168–214.

[7] William Saunders Hallaran, *An Enquiry into the Causes Producing the Extraordinary Addition to the Number of Insane* (Cork, 1810), 12.

[8] Ibid. 13.

[9] William Saunders Hallaran, *Practical Observations on the Causes and Cure of Insanity* (Cork, 1818), 24–5.

[10] Marianne Elliott, *Partners in Revolution: The United Irishmen and France* (New Haven, 1982), p. xiii.

for the English physician, anti-Jacobin fears lead to a playing up of the notion of infection, for Hallaran, in a society where opposition to English rule was capable of commanding immediate, wide sympathy, similar fears conjured images of a systemic instability.

Flexibility is also evident in the ways non-medical writers yoked male insanity with the subject of revolution. Perhaps most famously, Shelley enlisted images of male insanity into his poetry in such a way that madness attached to social oppression rather than to revolution: the maniac in *Julian and Maddalo* (composed in 1818) is 'a nerve o'er which do creep | The else unfelt oppressions of this earth' (lines 449–50). Revolution would be his cure.[11] Rousseau's insanity was a staple theme of anti-Jacobin polemicists, but it was also harnessed by writers of very different political persuasions. In *Fleetwood* (1805), William Godwin has MacNeil concede Rousseau's insanity grudgingly, as if the extraordinary 'resources of his own mind' were sufficient excuse for his paranoid self-absorption: 'It was difficult to persuade myself that the person I saw [in his moments of sublime inspiration] was the same as at others was beset with such horrible moments.'[12] Byron's depiction of Rousseau 'phrenzied', in the third canto of *Childe Harold's Pilgrimage* (1816), by comparison, hovers ambiguously between the madness of genius and a less attractive insanity: his mind raging 'with fury strange and blind' against 'self-sought foes', 'he knew | How to make madness beautiful'.[13]

To turn to the representation of madwomen in connection with political insurrection is to find in the main a less supple set of ideas, both within medical literature and more widely. Women occupied a special place within medical commentaries on revolution and insanity. When Burrows cites Rush on the susceptibility of populations to mental diseases in times of political upheaval, he recycles Rush's notes on the mysterious diseases and deaths of soldiers from mental causes, but like the American doctor he sees

[11] Timothy Clark, *Embodying Revolution: The Figure of the Poet in Shelley* (Oxford, 1989), 37.

[12] William Godwin, *Fleetwood* (1805), as cited and discussed in Edward Duffy, *Rousseau in England: The Context for Shelley's Critique of the Enlightenment* (Berkeley, 1979), 51.

[13] Lord Byron, *Childe Harold's Pilgrimage*, Canto III, lines 757, 756, 752, 729–30, in Jerome J. McGann (ed.), *Lord Byron: The Complete Poetical Works*, 7 vols. (Oxford, 1980–93), ii. 105–6. Duffy, *Rousseau in England*, 26–7, 30–1, 34, and esp. 37–45.

the effects of rebellion on the women as more striking and more 'natural'—the inevitable result of their nervous impressionability and cyclical 'derangement': 'The scenes that were passing suspended in the women hysterical and different complaints, and produced many others. Similar effects on the female sex were observed during the rebellion in Scotland, in 1745. The siege of Paris by the Allies in 1814 occasioned the female inhabitants much irregularity in the menstrual flux; and apoplexy and mania were more generally frequent.'[14] Although French writers and painters had adeptly exploited positive images of women in the early years of the Revolution, the period following the establishment of the new regime saw a distinct backlash against women in the public sphere, with the rhetoric of insanity increasingly used to pathologize and debar women's active intervention in political life.[15] In England, too, as Burrows recognized, the link between disorder in the body politic and disorder in the female body was likely to prove persuasive for a society taught to recognize, respect, and wherever possible soothe the nervous condition of femininity. The yoking of revolution with female irrationality was, in other words, symbolic but, at the same time, entirely convincing—and accordingly less tractable—for early nineteenth-century writers.

Moreover, where the male maniac was largely detachable from conventional associations and able to be turned to a variety of rhetorical purposes, the madwoman's sentimental associations were still strong enough in the 1790s and early 1800s to restrict what could be said through her. Anti-Jacobin writers seem to have found little to their purposes in a narrative which had traditionally aestheticized female insanity and ended, if it ended at all, in the

---

[14] Burrows, *Commentaries*, 20–1.

[15] See Joan B. Landes, *Women and the Public Sphere in the Age of the French Revolution* (Ithaca, NY, 1988), esp. 152–200; Jane Arbray, 'Feminism in the French Revolution', *American Historical Review*, 80 (1975), 43–62, esp. 51 on the commital of the feminist activist Théroigne de Méricourt to an asylum, and 57 on claims that women activists were generating public hysteria. Also William H. Sewell, Jr., 'Le Citoyen/la citoyenne: Activity, Passivity and the Revolutionary Concept of Citizenship', in C. Lucas (ed.), *The French Revolution and the Creation of Modern Political Culture*, ii: *The Political Culture of the French Revolution* (Oxford, 1988), esp. 118–19; Annette Rosa, *Citoyennes: Les Femmes et la Révolution française* (Paris, 1988), 218–24. For an important corrective to this emphasis, see Dorinda Outram, *The Body and the French Revolution: Sex, Class and Political Culture* (New Haven, 1989), 127–8.

sufferer's death. Burke's writing about madwomen and revolution draws strongly upon the furies of classical literature, but not at all on more sympathetic literary models of female insanity. His account of the revolutionary crowd at Versailles is one of the most widely quoted passages from the *Reflections*, depicting the summary execution of two gentlemen from the King's bodyguard amid 'the horrid yells, and shrilling screams, and frantic dances, and infamous contumelies, and all the unutterable abominations of the furies of hell, in the abused shape of the vilest of women'.[16] There is no place for sentiment within such a scene.

On the other hand, love-mad women seem to have offered little to support positive representations of revolution during the 1790s and early 1800s. The first pro-revolutionary novels are remarkable for the unshakeable sanity of their heroines. Thomas Holcroft, credited with being the first English novelist of the Revolution, was a friend of many of the English radicals, including Paine, Godwin, and Mary Wollstonecraft.[17] The heroines of his *Anna St Ives* (1792) and *The Adventures of Hugh Trevor* (1794–7) are resolutely clear-headed under romantic adversity, and, in their political acumen and moral stamina, they embody the most glowing hopes of the Jacobins. While rarely if ever criticizing conventional depictions of women's insanity as explicitly as Wollstonecraft did in *Maria*,[18] Holcroft and his followers evidently shared that novel's opposition to the sentimentalization of insanity. Anti-Jacobin novelists on occasion responded by creating recognizably derivative young women and punishing them with madness,[19] but even the most conservative writers were usually at pains to avoid rendering rev-

[16] Burke, *Reflections on the Revolution in France*, 72. Outram notes the gradual replacement of the term 'furies' with the medicalized concept of hysteria during the 19th cent.; *The Body and the French Revolution*, 128.

[17] Holcroft's *Anna St Ives* is described by J. M. S. Tompkins as the 'first full-blown revolutionary novel'—*The Popular Novel in England 1770–1800* (London, 1961), 300; see also R. M. Baine, *Thomas Holcroft and the Revolutionary Novel* (Athens, Ga., 1965).

[18] Discussed in Ch. 1.

[19] In Elizabeth Hamilton's *Memoirs of Modern Philosophers* (Bath, 1800), for example (a burlesque of Mary Hays, *The Memoirs of Emma Courtenay* (London, 1796)), Julia Delmond is seduced and intellectually corrupted by a Jacobin villain and abandoned. Driven insane by the recognition of her guilt, she recovers only long enough to point the moral: 'What palpitations, what terrors laid hold of my distracted mind! 'Twas then, then I first suspected the possibility of my having cherished false opinions' (ii. 307). On the tensions between sentimentalism and radicalism in Hays's novel, see Allene Gregory, *The French Revolution and the English Novel* (New York, 1915), 223–5.

olutionary women sympathetic through insanity. The beautiful heroine of Charles Lucas's *The Infernal Quixote: A Tale of the Day* (1801), for example, is sadly corrupted by a lover who encourages her to read Wollstonecraft, Godwin, Paine, Diderot, and Voltaire. She runs away with him to Paris, where he fails to keep his promise of marriage and rapidly proves unfaithful. Madness is not her fate, however. Having forfeited her self-esteem, she learns to act and think more soberly, and the novel rewards her with the love of a good man.

From the second decade of the nineteenth century, the general tenor of British politics changed, however, and with it the kinds of fiction being written on the subject. Gary Kelly has described the 1810s as characterized by a 'developing sense of crisis in national morality and leadership' arising from 'Britain's involvement in the Peninsular War', from 'the accession of the notoriously courtly and effeminate Prince of Wales to the powers of Regent', and from the rapid worsening of 'post-war economic distress and social unrest'.[20] Discontent was particularly feared in Ireland, after the failed rising of 1798 and the 1800 Act of Union; and in Scotland, where radical agitation against unemployment and poverty gave cause for serious concern, Walter Scott was far from alone in fearing an outbreak of civil war.[21] At the same time, these were years which offered little encouragement to revolutionary idealism. By 1814–15 the old monarchies had been restored to power in Europe. In England the Tories were clamping down on popular insurgency with the full weight of the law, and oppressive policy was the order of the day in Ireland and Scotland. It is in this context of considerable popular unrest, yet also of stringent government controls on its expression, that the convention of the love-mad woman can first be found regularly enlisted in stories of revolution.

The themes of love-madness and war were clearly coming together in fiction of the early 1810s, albeit often in little more than an attempt to enliven otherwise jaded material. Lady Charlotte Bury's *Self-Indulgence: A Tale of the Nineteenth Century* (1812) and Louis Sidney Stanhope's *Madelina: A Tale, Founded on Facts*

---

[20] Gary Kelly, 'Revolutionary and Romantic Feminism: Women, Writing, and Cultural Revolution', in Hanley and Selden (eds.), *Revolution and English Romanticism*, 126.

[21] See Sir Walter Scott, *The Visionary*, ed. Peter Garside (Cardiff, 1984), 5; and see Caroline Franklin, 'Feud and Faction in *The Bride of Lammermoor*', *Scottish Literary Journal*, 14 (1987), 18–31.

(1814) both fended off evident anxieties about the freshness of the convention by placing their love-mad women against the background of contemporary Anglo-French politics. Of the two, *Madelina* is perhaps the more remarkable for the sheer anxiety it conveys in relation to love-madness. The main narrative continually rebukes its heroine's tendency to hysterical breakdown over her misfortunes in love; yet the novel advances no less than four subplots about women driven insane by the loss of their lovers. *Madelina* is set during the Napoleonic Wars, but the implications of its conjunction of the themes of war and madness are barely tested. For a more extended examination of the love-mad woman's potentially symbolic relationship to political violence, it is necessary to turn away from English fiction to Anglo-Irish and Scottish novels of the 1810s, where stories about love-mad women were quickly becoming a vehicle for imagining the possibility of rebellion against English rule.

The admission of violence to the old sentimental and romance narrative had specifically literary implications: it allowed the convention to be recycled and renewed at a time when it had become somewhat stale and a regular target for criticism (see Chapter 2). The attraction of the love-madness convention for writers interested in the subject of rebellion is also not hard to see. The narrative's old sentimental associations and its more recent discrediting by anti-sentimentalists made it particularly well adapted to articulating both the romantic aspect of revolution and the force of opposition to it. This is not to say that the madwoman is an inherently conservative figure; rather, she provides a controllable narrative framework for thinking about revolutionary politics in a highly unstable political climate. She expresses both a fascination with rebellion and a more sober recognition of its cost, within narratives that are sometimes in agreement with conservatism and sometimes (more complexly) in sympathy with radical politics but unable or unwilling to imagine their success in Britain in the 1810s, unless it be in the form of a self-consuming act of political revenge.

Charles Maturin's *The Milesian Chief* was published in 1812, to a mixed critical reception.[22] Despite the title, *The Milesian Chief* is

---

[22] See e.g. *Westminster and Quarterly Review*, 46 (1847), 617–18. The novel enjoyed some success in France. Balzac published a translation of *The Milesian Chief*

principally the story of a woman, Armida Fitzalban—the name suggesting that Maturin, like many of his Romantic contemporaries, was an admirer of Torquato Tasso's *Gerusalemme Liberata* (1581).[23] In Tasso's epic poem, Armida, a beautiful Damascan sorceress, disrupts the First Crusade on its way to Jerusalem, waylaying many of the Christian camp and doing some damage to the cause through her seduction of Rinaldo. The poem's reputation had been kept alive by eighteenth-century opera,[24] and its thematic interweaving of war and destructive sexual passion made it a highly charged precursor for much early nineteenth-century Romantic writing on women and revolution. *Gerusalemme Liberata* had several closer adaptations than *The Milesian Chief* during this period, the best known being Shelley's 'Laon and Cythna' (1817). Shelley's heroine, like Maturin's, dies along with her lover when the revolt is suppressed, but 'Laon and Cythna' might have found much in Maturin's novel to feed its despair of revolution; and little to encourage its hopes for the revival of radicalism. Indeed, Shelley and Maturin represent opposed positions within Irish politics in the 1810s. February 1812 saw Shelley attending a meeting of the United Irish Society and speaking in support of the republican movement;[25] Maturin, a Protestant minister and a descendant of French Huguenots, was, *pace* Coleridge, by far the more conserva-·tive figure.[26]

The *Milesian Chief* was Maturin's third novel. As in all his fiction, the plot is complex in the extreme.[27] Armida Fitzalban

in 1828: *Connal; ou, Les Milesians,* par Maturin, de *Melmoth le voyageur,* etc., traduit de l'Anglais par Mme La Comtesse [Mole] (Paris, 1828). Balzac's own satiric sequel, *Melmoth réconcilié,* followed in 1835.

[23] On Tasso's popularity among the Romantic writers between 1750 and 1850, and the particular English fascination with his madness, see Clark, *Embodying Revolution,* 178–84.

[24] Esp. Handel's *Rinaldo* (1711), which was a sensational success in England. See Stanley Sadie (ed.), *The New Grove Dictionary of Music and Musicians,* 20 vols. (London, 1980), viii. 86.

[25] See Kenneth Neil Cameron, *The Young Shelley: Genesis of a Radical* (New York, 1950), 147; and Michael Henry Scrivener, *Radical Shelley: The Philosophical Anarchism and Utopian Thought of Percy Bysshe Shelley* (Princeton, 1982), 56.

[26] For Coleridge's attack on Maturin's *Bertram* on the grounds of its Jacobinism, see the *Biographia Literaria,* ch. 23, in Kathleen Coburn (gen. ed.), *The Collected Works of Samuel Taylor Coleridge* (Princeton, 1976–), vii, ed. James Engell and W. Jackson Bate.

[27] For a more detailed résumé of the plot, see Dale Kramer, *Charles Robert Maturin* (New York, 1973), 44–53.

is half-Italian, half-English. Notwithstanding her air of 'Italian indolence', she is mistress of a formidable array of talents and the first chapters of the novel are primarily a catalogue of her powers of artistic performance—acting, singing, lute-playing, and recitation of poetry (her own translations from the Persian). Soon after the start of the novel, Armida is taken back to Ireland by her melancholic and oddly capricious father. During a brief stay in London *en route* to Ireland, she is persuaded to engage herself to Colonel Wandesford, who goes to take up a commission with the British forces in Ireland. But by the time her fiancé arrives in the country Armida has lost her heart to her rebel cousin, Connal O'Morven. Armida's strictly Calvinist father disowned his sister after her marriage into the O'Morven family. She and her husband are now dead, and her sons and father-in-law are bankrupt, but, unlike his brother, Connal has refused to betray their ancient Milesian heritage by joining the English army. Instead, he leads a band of Irish rebels, plotting the overthrow of English rule; but his commitment to militant politics rapidly weakens when he meets Armida. Captivated by her, he undertakes to persuade his men that they must give up rebellion and throw themselves on the mercy of the English government, offering himself as a hostage if necessary. His chances of success are thwarted by the interruption of Colonel Wandesford. Connal prevents his men from killing Wandesford, in return for a promise that the Colonel will intercede for them with the government. But Wandesford, angered by the attachment which has grown up between his fiancée and the Milesian rebel, breaks his word. The following night, he leads a raid on the O'Morven tower, and burns it to the ground as a warning to the insurgents. From this point, Connal has no choice but to take up the rebel cause.

When the rebellion is suppressed, and false reports circulate that Connal has been killed, Armida's fragile mental health gives way completely. She returns to sanity only long enough to take a slow-acting dose of poison when her stepmother forces her to agree to a marriage with Connal's brother Desmond. As she lies dying, Connal (briefly alive after all) is taken out before the firing-squad and shot. His brother, Desmond, hurls himself in front of the bullets so that both men die together. Desmond's wife, already deranged by the death of their child and by a long enforced separation from Desmond, goes raving mad and dies by her husband's side. Lastly Armida drags herself to the heap of corpses so that she, too, can expire in the arms of her dead lover.

Although Maturin persists in referring to his heroine as English, it is, of course, Ireland that forms the context for this story of self-destructive romantic passion. Equally, romance fiction provides the controlling framework for imagining Irish rebellion against English rule. In *The Milesian Chief*, illicit sexual passion is both a metaphor for political insurgency and the reason that insurrection is ultimately thwarted. Connal is torn between the two forms of 'insanity': the woman's mad and maddening passion, on the one hand, and the revolutionary madness of his grandfather on the other.

The complexity of that analogy is suggested by a scene late in the novel, when Armida is kidnapped by one of Connal's subordinates. Brennan delivers her over to Connal's grandfather, and she narrowly escapes being murdered by the old man, who is convinced that the liberation of Ireland from English rule is secured if he kills her. His desire for political vengeance, under Brennan's goading, takes on the language and the gestures of rape, though the old man is too feeble to make it effective:

'Traitress!' he exclaimed, though speaking English with difficulty; 'traitress! you are in my power at last: you shall at last feel it. I know you well, though you assume that appearance of youth and beauty to deceive me. You are the Queen of England: the false daughter of the heretic Henry. You have dispossessed me of my rightful dominion . . .' . . . Twice the old man, grasping the bayonet which Brennan forced into his hand, aimed it at Armida's breast: with the rage of a monster he then attempted to tear off the thin covering of her bosom. (iii. 157–8, 159–60)

Armida has heard something of the old man's story from a female servant. O'Morven has been deranged since the night the O'Morven tower was burnt to the ground, though poverty and grief had evidently undermined his reason long before:

what with hardship, and grief, and wandering, his senses are quite gone, and he thinks nothing but the life of one of your family could restore him to his reason. And sometimes he takes you for Queen Elizabeth, who he says was the first cause of the Protestant heresy, and says you came over from Italy to invade his country, and take his lands and castle from him; for he mixes up things so in his madness: and sometimes he will wander about at night, crying to them to bring him back to his castle, or to tell him who took away his understanding from him . . .
(iii. 144–5)

The madness of the old Irishman differs critically from that of Armida later, for, where her derangement results from the supposed death of her lover, his is caused by the destruction of the castle. He looks for the joint restoration of his property and his sanity; 'to bring him back to his castle' is to bring him back to his reason; and the burning of the tower takes on a strong symbolic as well as causal link with the destruction of his mind.[28] Women's madness never has such a correlation with the loss of property, or only in so far as women's property is seen to consist in an emotional claim. Women go mad when they lose a lover or a child in Maturin's novel, as everywhere else in eighteenth- and early nineteenth-century fiction. Men go mad when they lose their land, their money, or their status.

Connal's grandfather nevertheless blames his distress on women. Ireland's political subordination to England and its partial religious conversion are summed up for him in the figure of Elizabeth I and, by association, in Armida, the daughter of the Englishman who usurped the old man's lands. There is little in the novel to suggest that Maturin was concerned to correct the resulting displacement on to women of the symbolic responsibility for violence. The language of love shifts all too easily into the language of political conflict in *The Milesian Chief*, so that what Connal's brother terms 'the sweet madness of [women's] influence' readily becomes the bitter madness of their political influence. Brennan's probing interrogation of Armida when she is kidnapped encapsulates a continual pathologization of her, reinforced more covertly throughout the novel by the narrative voice and by the trajectory of the plot: ' "Do you deprecate murder, and mischief, and horror then, gentle lady?" said Brennan, in a voice of artificial softness: "why do you cause it then? Why do you?" he repeated, bending over' (iii. 153). The passage is the most genuinely sexually threatening in a novel which has its full complement of scenes of sexual predation, and Brennan's words sum up an attitude that pervades Maturin's text. Men's private relations with women seem to offer a possible field for reasserting authority when it has been lost everywhere else, as the encounter between Armida and Connal's grandfather makes clear. Women's madness from love and the Irish people's rebellion against economic hardship thus increasingly mirror each other,

---

[28] Henry Mackenzie's *The Man of Feeling* (1771) presents the most famous novelistic expression of the old belief that the brains of the mad burned with fever.

becoming one another's metaphors—but just how far Maturin intended to romanticize the rebel cause in the process is difficult to ascertain, given his extreme unwillingness to portray Connal as an insurgent. John Harris emphasizes that Maturin was anything but a rebel in his political sympathies, and *The Milesian Chief* is certainly no positive portrayal of insurgency.[29]

In part, this is because, like much of Maturin's fiction, it is paranoiacally anti-Catholic, although it does allow something of an alternative political standpoint to be seen. *The Milesian Chief* remains passionately nationalist within that vein of nationalism so often found in early nineteenth-century Anglo-Irish fiction, dwelling nostalgically on stories of Ireland's glorious past, her folkloric and druidical history—all combined with intense anger at Ireland's present state: 'The country is bleeding under ignorance, poverty, and superstition, and we cast over its wounds a gay embroidered garment of voluptuousness, beneath which the heavings and shudderings of its agony are but more frightfully visible' (ii. 41). Recalling the 'voluptuousness' with which Armida hides her unhappiness in the opening chapters of the book, and the (all too many) 'heavings and shudderings' of romantic agony she undergoes in Ireland, the passage seems to offer her as a symbol for the political sufferings of the nation. But that symbolism is necessarily conflicted and awkward. Although, at some points, Armida represents Anglo-Irishness, in danger of being ravaged by the rebels, elsewhere she is compellingly, if inconsistently, imagined through rebel eyes, aligned with Elizabeth I as the invading English presence—the cause of Ireland's bitter madness. In the story of this woman, captivated by the rebel leader, doing much to placate his political resentment, but finally driven to madness and suicide by his death in the rebel cause, Maturin seems to have found the corrective lens he needed through which to view a revolutionary drama he was unable to resist imagining, but equally unwilling to endorse.

Something of the same desire both to entertain and to disown the spectacle of rebellion is evident in the most notorious novel of the 1810s to share Maturin's interest in the madwoman as a symbol of Irish rebellion. Lady Caroline Lamb's *Glenarvon* (1816) was a kiss-

---

[29] John B. Harris, *Charles Robert Maturin: The Forgotten Imitator* (New York, 1980), 10–11.

and-tell account of her affair with Lord Byron—or as Byron put it more ruefully, a case of '—— and publish'.[30] It depicts Byron as the child-murdering Glenarvon, who seduces Lady Calantha, a thinly disguised Lady Caroline, then abandons her to madness and death (it is a plot which reads more poignantly in the light of Lady Caroline's history of mental suffering and her early death). 'Madness' functions largely as an exculpatory term in Lamb's vocabulary: Calantha's love for Glenarvon becomes 'a mad and guilty attachment' rather than a merely guilty one, a 'maddening disease' (ii. 146, 156). But even in this hugely self-pitying *roman-à-clef*, love-madness is highly politicized.

Glenarvon is more than a child-murderer and callous seducer of women. He leads a group of political insurgents in Ireland who, like Connal's men in *The Milesian Chief*, seek to deliver Ireland from English rule. A man of 'splendid genius and uncommon faculties' Glenarvon's principal effect on those around him is to drive them to madness of one kind or another. Scarcely has he arrived in Ireland when a 'singular and terrific inmate' is discovered in his retinue: 'a maniac! who was however welcomed in with the rest of the strange assemblage, and a room immediately allotted for his reception . . . Even in his most dreadful paroxysms, when all others were afraid of approaching him, Glenarvon would calmly enter into his chamber, would hear his threats unawed,—would gaze on him, as if it gave him delight to watch the violence of misguided passion; to hear the hollow laugh of ideotsy, or fix the convulsed eye of raving insanity' (ii. 89). Glenarvon brings the same gratified delight to the sight of the three women maddened by love for him in this novel; and to the similarly 'misguided' passion he provokes in the Irish people (ii. 91). The thematic interconnection is clear in *Glenarvon*, where seduction and politics continually overlap. When Glenarvon plays his mistress's harp, or sings 'the minstrelsy of the bards of other times', he 'inspire[s] the passions which he fe[els] and inflame[s] the imagination of his hearers to deeds of madness' (ii. 92). When he wishes to present Calantha with a love-token, he chooses an emerald ring, engraved with the armorial bearing of Ireland, a harp. 'I mean it merely politically', he teases her (ii. 158).

---

[30] Letter to John Murray, quoted in James L. Ruff, introduction to fac. edn. of Lady Caroline Lamb, *Glenarvon* (New York, 1972), p. ix.

As the representative of the English government in Ireland, Calantha's husband, Lord Avondale, thus has the dual task of quashing political rebellion in the country and of subduing a private rebellion on the part of his wife—both provoked by the one man. But where, in *The Milesian Chief*, the woman's link with political insurgency was primarily symbolic (and uncomfortably so), in *Glenarvon* the connection is a literal one. At this point, however, Calantha has a surrogate. The novel depicts, in parallel with her, another victim of Glenarvon's seductions: a nun named Elinor St Clare,[31] whose actions also draw closely on Lady Caroline's life at several points. Calantha collapses and—eventually—dies under the misery of desertion and the awareness of guilt, but St Clare is not so fragile. When Glenarvon seduces her, she cross-dresses as a boy and lives with him openly as a mistress.[32] When he abandons her, St Clare, in her madness, reappears in soldier's uniform, marching by the side of a dejected young man who has lost his lover to Glenarvon. St Clare becomes the figurehead and co-leader of Glenarvon's rebel party; and where she would formerly have testified in her madness to the disappointment of her love, here disappointment and a desire for revenge are channelled entirely into politics. Glenarvon's parting words to his men read tamely alongside her far more militant speech, and when the Irish men find that Glenarvon has betrayed them politically much as he betrayed their women emotionally, she takes the men to war without him. Near the end of the novel, she addresses the remaining rebels, spurring them on in their struggle:

'It will be a dreadful spectacle to see the slaughter that shall follow,' said St. Clare. 'Brothers and fathers shall fight against each other. The gathering storm has burst from within: it shall overwhelm the land. . . . What though

---

[31] The name suggests a debt to Lady Morgan's Anglo-Irish novel of 1803, *St Clair; or, The Heiress of Desmond*, in which St Clair and his lover are destroyed by uncritical devotion to the principles of Rousseau. The melancholic St Clair dies when his plot to elope with the heiress is foiled; the heiress herself declines into insanity and expires, having recovered sanity only long enough to regret that she allowed passion to pervert her reason. There may also be a debt to Maturin's *The Wild Irish Boy* (1808) in which a female St Clair solicits the hero's life story in letters but dies of excessive passion when she reads them.

[32] Lamb's cross-dressing exploits during her affair with Byron are well known. She claimed also to have dressed as a page-boy while writing *Glenarvon*, secretly, at night; and to have presented herself to the publisher's copyist as a 14-year-old serving-boy. For discussion, see Peter W. Graham, *Don Juan and Regency England* (Charlottesville, Va., 1990), 95.

in vain we struggle . . . the spirit of liberty once flourished at least; and every name that perishes in its cause shall stand emblazoned in eternal splendour—glorious in brightness, though not immortal in success.' (iii. 287–9)

As Glenarvon's castle goes up on flames, Elinor and her horse plunge to their death from the clifftop; but her revenge is completed posthumously. Glenarvon, attempting to flee to Holland by ship after wounding Lord Avondale in a duel,[33] is persecuted by the mournful ghosts of the women he drove to madness. His sanity gives way in the closing pages and, as he dies, a voice from the deeps assures the reader that hell is his destination.

In *Glenarvon*, both illicit love and political insurgency are ultimately self-destructive. But, like Maturin, Lamb seems to have recognized in the love-mad woman a figure capable of bringing the required ambiguities of tone to the subject of rebellion. For all her rebelliousness, Lamb was very much a product of the aristocratic Whig circles in which she moved,[34] and the politics of the novel are less subversive than her reputation would suggest. In the increasingly anti-sentimental climate of the 1810s, love-mad women were clearly romantic but wrong, and the resulting ambivalence attaching to Elinor St Clare's involvement in nationalist militancy seems to have suited Lamb's purposes. In *Glenarvon*, popular insurrection forms a backdrop for romance and an extension of the love-mad woman's repertoire rather than a subject in its own right. Any more serious contemplation of the connections between Romanticism and nationalism is left to the reader.

Sir Walter Scott is known to have read *Glenarvon*, and it is likely, though by no means certain, that he had also read *The Milesian Chief*.[35] Whatever his feelings about the particular merits of either work, their influence on his fiction seems clear. Between 1818 and 1820 Scott published three novels, all of which dealt with love-madness in contexts of social and political unrest. In *The Heart of Midlothian* (1818) the symbolic tie between love-madness and re-

[33] In the first 1816 edition the wound is superficial, in the second it is fatal.

[34] See Graham, *Don Juan*, 94.

[35] On *Glenarvon*, see Lady Abercorn's letter to Scott, in *The Letters of Sir Walter Scott*, 12 vols., ed. H. J. C. Grierson *et al.* (London, 1932–7), iv. 364 n.; and, more obliquely, Scott's comment in 1825, 'I hear Calantha is loose again', ix. 246. There is no mention of *The Milesian Chief* in Scott's *Letters*, though most of Maturin's other works are commented on soon after publication.

bellion has one of its most memorable formulations. George Staunton leads the storming of Edinburgh's Tolbooth Prison disguised in the borrowed clothes of Madge Wildfire, who has been insane ever since he abandoned her with child. In that act of cross-dressing is summed up the displacement of political insurgency on to the figure of the madwoman, as her private, familiar tragedy becomes the costume for a political drama enacted by men.

Ivanhoe (1820) is, if anything, more explicit in its yoking of female insanity and insurgency. Early in this tale of Saxon–Norman rivalry in the late twelfth century, the 'fair Jewess' Rebecca is imprisoned in a turret room in the castle of the villainous Front-de-Bœuf. Rebecca finds herself

in the presence of an old sibyl, who kept murmuring to herself a Saxon rhyme, as if to beat time to the revolving dance which her spindle was performing upon the floor. The hag raised her head as Rebecca entered, and scowled at the fair Jewess with the malignant envy with which old age and ugliness, when united with evil conditions, are apt to look upon youth and beauty.[36]

The stories of Rebecca and of Ulfried, 'the old crone of the turret', are deliberately counterpointed in *Ivanhoe*. Threatened with rape by one of Front-de-Bœuf's companions, Rebecca runs to the parapet and prepares to throw herself from the battlements rather than submit to violation. Impressed with her virtue and ashamed of his own villainy, the knight repents and leaves her. Faced with the same choice between chastity and death years earlier, Ulfried was not so courageous. As a young girl, she was forced to watch while Front-de-Bœuf murdered her Saxon father, before herself being abducted and raped. Lacking the courage to commit suicide, she lived with the Norman as his mistress and contented herself with petty retaliation, dabbling in witchcraft, promoting dissension and bloodshed among her persecutors, and going slowly crazy.

In many respects, *Ivanhoe* overturns the determining features of the love-madness convention, rewriting them in colours of violence. Ulfried was raped, not seduced, and her madness was born as much of hatred as of love; yet her mania still preserves the crucial link with desertion by a man. Only when Front-de-Bœuf abandoned her for younger women, Ulfried claims, did she become 'a hag': 'condemned to hear from my lonely turret the sounds of revelry in

---

[36] Sir Walter Scott, *Ivanhoe*, ed. A. N. Wilson (Harmondsworth, 1982), 244.

which I once partook, or the shrieks and groans of new victims of oppression' (278). Her insanity is, in her own words, tainted by the 'deep, black, damning guilt' of complicity with the enemy (276).

In the course of *Ivanhoe*, however, Ulfried is reunited with her Saxon origins and repairs the cowardice of her youth. When the Saxon army besieges the castle walls, she assists her countrymen from within, setting fire to the castle and incinerating her persecutor in his chamber. As the fire spreads, Ulfried appears on the battlements, exulting maniacally in her victory. In that moment of insane triumph, Scott grants her once again her Saxon name of Ulrica:

The fire was spreading rapidly through all parts of the castle, when Ulrica, who had first kindled it, appeared on a turret, in the guise of one of the ancient furies, yelling forth a war-song, such as was of yore raised on the field of battle by the scalds of the yet heathen Saxons. Her long dishevelled grey hair flew back from her uncovered head; the inebriating delight of gratified vengeance contended in her eyes with the fire of insanity; and she brandished the distaff which she held in her hand, as if she had been one of the Fatal Sisters, who spin and abridge the thread of human life. . . . The maniac figure of the Saxon Ulrica was for a long time visible on the lofty stand she had chosen, tossing her arms abroad with wild exultation, as if she reigned empress of the conflagration which she had raised. At length, with a terrific crash, the whole turret gave way, and she perished in the flames which had consumed her tyrant.   (343–6)

This is the madwoman's crowning moment as a political insurgent in early nineteenth-century fiction, but beneath the melodramatic flourish, there is a degree of caution implicit in the use of the madwoman as an agent of war. When 'gratified vengeance' contends with insanity in Ulrica's eyes, Scott comes very close to eliding the two. In effect, madness both justifies revenge and determines its limits. The fire which spreads around Ulrica blazes like the 'fire of insanity' in her eyes, so that her actions appear to express an organic capacity for retribution, an unmediated anger which wreaks disaster by the sheer projection of anger. Henry Mackenzie's sentimental madwoman complained that her brain was dry 'and it burns, it burns, it burns'. Ulrica turns that metaphorical, symptomatic burning of the maniac outward, literalizing it and inflicting it upon her enemies so that they burn with her. Such elemental vengeance takes on a mythic dimension in Scott's writing.

No longer just an old crone spinning in a turret, Ulfried finally claims her kinship with 'the ancient furies' and 'the Fatal Sisters, who spin and abridge the thread of human life'.

In all Scott's writing about love-mad women, however, one novel stood out, at least by his own estimation: 'of all the murders that I have committed in that way, and few men have been guilty of more, there is none that went so much to my heart as the poor Bride of Lammermoor'.[37] Many of Scott's readers agreed with his judgement. Published in 1819, between *The Heart of Midlothian* and *Ivanhoe*, *The Bride of Lammermoor* rapidly became the century's most famous depiction of love-madness. For many readers this was *the* tragic novel of the age, its reputation even persuading the Queen to venture into reading fiction for the first time in 1838.[38] Perhaps the most remarkable sign of the story's impact on its first audiences was its remarkable generative power. Between 1819 and 1908 it inspired more than twenty-five stage plays, melodramas, and operas, in over 500 productions in France, Germany, Denmark, Italy, and America, as well as in Britain.[39] Only one was really successful: Donizetti's *Lucia di Lammermoor*. First performed in Naples in 1835, *Lucia* established Scott's bride as the most famous of all opera's madwomen, and its recension of the novel has significantly coloured the book's reputation ever since.

The opera tells the quintessential story of a woman deranged by love. Lucy/Lucia Ashton contracts a secret engagement with an impoverished nobleman whose family lands and castle her father has taken over in a successful series of lawsuits. Prevented by her family from marrying Edgar/Edgardo, Lucy is forced into a marriage with another man. On her wedding-night she goes raving mad and stabs her husband. After an extended cabaletta, she collapses on stage and dies. Jerome Mitchell notes the principal changes made in adapting the novel for opera: 'No Scottish local color is

[37] John Gibson Lockhart, *Memoirs of the Life of Sir Walter Scott, Bart.*, 2nd edn., 10 vols. (Edinburgh, 1839), x. 191.

[38] See *The Girlhood of Queen Victoria: A Selection from Her Majesty's Diaries between the Years 1832 and 1840. Published by the Authority of His Majesty the King*, ed. Viscount Esher, 2 vols. (London, 1912), i. 260.

[39] H. Philip Bolton, *Scott Dramatized* (London, 1992), 297. See also Jerome Mitchell, *The Walter Scott Operas: An Analysis of Operas Based on the Works of Sir Walter Scott* (Alabama, 1977), and Henry Adelbert White, *Sir Walter Scott's Novels on the Stage* (New Haven, 1927), 74–101.

manifest in either music or text' of Donizetti's *Lucia*.[40] Numerous characters essential to the novel are also omitted, among them Lord and Lady Ashton, Caleb Balderstone (Ravenswood's faithful retainer), old Alice, and the three witches. As Mitchell sees it, the omissions enable the opera to avoid the weaknesses evident in earlier adaptations. Donizetti and Cammarano (his librettist) 'knew what they were about':[41]

What Cammarano and Donizetti have done instead is to concentrate on the love-story of Edgar and Lucy and to omit everything in the novel that is not obviously related to it. They capitalize on effective scenes already in the novel—the picturesque, quiet scene at the fountain where Edgar and Lucy plight their troths and the dramatic confrontation between Edgar and the signers of the marriage contract—and they create effective scenes from material that Scott does not fully develop: the meeting of Edgardo and Enrico at Wolf's Crag and especially the celebrated mad scene.[42]

The last point is crucial. The opera makes madness its most expansive dramatic and musical moment (an emphasis highlighted by the mad scene's popularity as a solo item for concerts in the later nineteenth century).[43] The ravings of the madwoman are melodramatic theatre, and the perfect vehicle for bel canto. *Lucia di Lammermoor* is a diva's opera. As Gary Schmidgall notes in *Literature as Opera* (1977), the operatic genre it belongs to 'was designed to serve and display' the singer: 'Not far below the surface of most bel canto heroines like Lucia and Maria [Stuarda] lies [a] psychology of artistic virtuosity and self-indulgence.'[44]

Scott's representation of the madness of Lucy Ashton is remarkable, by comparison, for its brevity. Readers who turn to the novel in expectation of the histrionic display Donizetti stages will be disappointed. This is all Scott gives us of the scene in which Lucy is found in the bridal chamber after stabbing the man she has been compelled to marry:

[40] Mitchell, *The Walter Scott Operas*, 144.
[41] Ibid.
[42] Ibid.
[43] Bolton, *Scott Dramatized*, 297.
[44] Gary Schmidgall, *Literature as Opera* (New York, 1977), 146, 147. For the current debate in literary theory over the gender politics of opera's madwomen as it relates specifically to Lucia, see Catherine Clément, *L'Opéra; ou, La Défaite des femmes*, trans. as *Opera or the Undoing of Women* by Betsy Wing (London, 1989), 88–90.

one of the company, holding his torch lower than the rest, discovered something white in the corner of the great old-fashioned chimney of the apartment. Here they found the unfortunate girl, seated, or rather couched like a hare upon its form—her head-gear dishevelled; her night-clothes torn and dabbled with blood,—her eyes glazed, and her features convulsed into a wild paroxysm of insanity. When she saw herself discovered, she gibbered, made mouths, and pointed at them with her bloody fingers, with the frantic gestures of an exulting demoniac.

Female assistance was now hastily summoned; the unhappy bride was overpowered, not without the use of some force. As they carried her over the threshold, she looked down, and uttered the only articulate words that she had yet spoken, saying, with a sort of grinning exultation,—'So, you have ta'en up your bonny bridegroom?'   (338)

For all its compactness, the scene is powerfully dramatic. Compared with her precursors in the nineteenth-century novel, Lucy Ashton is extraordinarily shocking. The 'unfortunate girl' is reduced to animality: 'couched like a hare upon its form'—a simile which identifies her closely with the novel's malevolent witches (according to popular tradition, witches were able to metamorphose into hares). Her language, too, up till now genteelly English, has become the broad Scots of the hags.[45] This is not the gentle erotomaniac of sentimental and romance fiction, but 'an exulting demoniac', and Scott stresses the element of gratified revenge. In a ghastly assertion of power, Lucy has turned what promised to be the unwanted penetration of her body into the violent penetration of her husband's. Bucklaw's blood, not her own, is spilt, and the staining of her night-clothes brutally parodies the consummation he and the wedding-guests looked for.[46] Her 'only articulate words' state the right she has claimed to spurn a choice of husband that her family, not she, has made: 'So, you have ta'en up your bonny bridegroom?'

For early reviewers of the novel, unprejudiced by Donizetti's opera, the brevity of the catastrophe was less surprising than the

[45] See Fiona Robertson's introduction to the World's Classics edition of *The Bride of Lammermoor* (Oxford, 1991), esp. p. xxiii on Lucy's 'desperate rebellion' as the culmination of a pattern of images of 'female rule and female violence'. According to Scott's introduction, Lady Ashton, too, was held by popular tradition to have been a witch.

[46] The point is underlined in Scott's 1830 introduction, where he notes the wedding-guests' reluctance to interfere when shrieks are heard from the bridal chamber because they assume the consummation is taking place.

degree of foreshadowing it received in the narrative. Readers of the 1830 revised edition reached the scene of Lucy's insanity in chapter 34 having literally heard it all before. A lengthy introduction, added by Scott in that year, not only outlined the plot of the novel, but recounted the events of the wedding-night in detail—almost the same detail he used later:

On opening the door, they found the bridegroom lying across the thresh-old, dreadfully wounded, and streaming with blood. The bride was then sought for: She was found in the corner of the large chimney, having no covering save her shift, and that dabbled in gore. There she sat grinning at them, mopping and mowing, as I heard the expression used; in a word, absolutely insane. The only words she spoke were, 'Tak up your bonny bridegroom.' (4)

The end of the novel places slightly greater emphasis on Lucy's torn and bloodied clothing, and on her association with the witches, but the substance and the language are very close indeed.

The 1830 introduction leaves the reader in no doubt of the nature of the tragedy to come, but even without that assistance, it was impossible to miss the insistent allusions to a malevolent fate hanging over the bride of Lammermoor. Omens are everywhere: a bull attacks Lucy Ashton and her father, just pages after Scott has reminded listeners that the bull's head is traditionally a token of death; a raven falls from the sky staining her dress with blood as she and Ravenswood plight their troth; and the story of the Naiad of the fountain, who dies when she is betrayed by her lover, symboli-cally seals the girl's fate. Most of the allusions point towards an undetermined form of tragic death, but at certain points, in a novel thick with allusions to *Hamlet*,[47] Scott invokes Ophelian madness more nearly. Lucy's passion for melancholy ballads is established even before she meets her lover, and one passage particularly presses the allusion home. With her 'soft and yielding' temper, Scott notes, Lucy will be 'borne along by the will of others, with as little power of opposition as the flower which is flung into a running stream' (40). Early readers of *The Bride of Lammermoor* were troubled by this obsessive anticipation of the novel's tragic ending. Nassau Senior reviewed the novel for the *Quarterly Review* and remarked on its 'deviation from the usual management of a narra-

---

[47] See Frank McCombie, 'Scott, *Hamlet*, and *The Bride of Lammermoor*', *Essays in Criticism*, 25 (1975), 419–36.

tive': 'The fatal nature of the catastrophe is vaguely indicated in the very beginning; at every rest in the story it is more and more pointedly designated; and long before the conclusion we are aware of the place and means of its accomplishment.' For this reader, the result was not a diminution of interest in the narrative but on the contrary an enhancement he felt obliged to account for: 'our interest in the story is strengthened, instead of being destroyed, by our fore-knowledge of the conclusion. How is this managed? How is that which generally deadens the reader's interest made, in this instance, its auxiliary?'[48] Nassau Senior considered both Lucy and Ravenswood seriously flawed as sympathetic individuals, and argued that the novel succeeded by virtue of its formal excellence rather than by the appeal of its characters (Scott famously admits that, had Ravenswood thought about Lucy in any depth, he would probably have found her insipid). The *Quarterly Review* concluded that *The Bride of Lammermoor* is able to compel our interest, even though we know the plot already, because it is the 'privilege of tragedy' that knowing in advance the fate awaiting characters should increase our sympathy for them.

Leslie Stephen's 1871 essay for the *Cornhill Magazine* summarized the opposing argument. He, too, concentrated on the fact that Scott evidently put narrative organization before the representation of sympathetic characters, and quoted Carlyle in support of his objections:

'While Shakespeare works from the heart outwards, Scott,' says Mr. Carlyle, 'works from the skin inwards, never getting near the heart of men.' . . . And though many good judges hold *The Bride of Lammermoor* to be Scott's best performance, in virtue of the loftier passions which animate the chief actors in the tragedy, we are, after all, called upon to sympathise rather with the gentleman of good family who can't ask his friends to dinner without an unworthy device to hide his poverty, than with the passionate lover whose mistress has her heart broken. Surely this is the vulgarest side of the story. Scott, in short, fails unmistakeably in pure passion of all kinds; and for that reason his heroes are for the most part mere wooden blocks to hang a story on.[49]

---

[48] Nassau Senior, unsigned review of eight Scott novels, in the *Quarterly Review* (Oct. 1821); repr. in John O. Hayden (ed.), *Scott: The Critical Heritage* (London, 1970).

[49] Leslie Stephen, unsigned article 'Some Words about Sir Walter Scott', *Cornhill Magazine* (Sept. 1871); repr. in Hayden (ed.), *Scott*.

Stephen accuses Scott's novel of 'vulgarity' for the same reason Nassau Senior judges it 'a tragedy of the highest order': both recognize that telling the story is somehow more important to Scott than the characters he describes. For Stephen, the concentration on plotting flaws the work irreparably; for Nassau Senior, in the end, it defines its excellence.

The distinction these nineteenth-century readers made between the slight emotional interest of Edgar Ravenswood and Lucy Ashton and the keen interest generated by the telling of their story goes to the heart of the novel's interest as a reworking of the madwoman convention. Stephen's sense that *The Bride of Lammermoor* was curiously uncommitted to 'the passionate lover whose mistress has her heart broken' is unfair to the emotional power of the novel, but it does capture the strangely detached quality of Scott's engagement with Lucy Ashton's tragedy. Of all Scott's characters, Lucy suffers most from this novel's 'intrusive literariness'—its insistent reference to the tragic burden of the past, whether it be through allusion to Shakespeare, or to popular ballads, or to common gossip.[50] The sense of overdetermination may, in part, be a result of Scott's ill health when he wrote the novel. He was obliged to dictate the last half of the work from his sick-bed, and could remember almost nothing about it afterward.[51] In such a state, it was more likely he would fall back on a known or conventional story than at other times. Yet a comment recorded by his biographer, Lockhart, strongly suggests that the peculiar quality of the love-madness convention was also part of the explanation. When Scott reflected on his own fascination with stories of love-mad women, he noted that the narrative seemed to him to possess a force which defied authorial intervention, leading him to remark of the last of his love-mad heroines, Clara Mowbray: 'I could not save her, poor thing—it is against the rules.'[52] That sense of an unchallengeability about the love-madness narrative continually colours Scott's treatment of Lucy Ashton's story, and it results in a shift of emphasis away from the details of the narrative *per se* in favour of the conditions shaping this particular recital of the familiar tale.

In this respect, Scott's treatment of the love-madness convention

---

[50] See Robertson, Introduction to *The Bride of Lammermoor*, pp. xix, xxvii.
[51] See Lockhart, *Memoirs of the Life of Sir Walter Scott*, vi. 88–90.
[52] Ibid. x. 191.

receives a great deal of support from a series of reflections on the story-telling traditions of the common people of Scotland. *The Bride of Lammermoor* signals its interest in folklore early on. The first edition was published anonymously and began with a lengthy diversion from the Ravenswood–Ashton story. Continuing a pattern set up in the early *Tales of my Landlord*, the novel is supposedly narrated by Peter Pattieson, a schoolmaster and collector of Scottish folklore, whose manuscripts are edited after his death by the pedantic Jedediah Cleishbotham. The opening chapter is principally devoted to Pattieson's recollections of the artist Dick Tinto, from whom he claims to have obtained the manuscript notes on which the novel is based. Ostensibly educational for the reader and a warning to the author 'against seeking happiness, in the celebrity which attaches itself to a successful cultivator of the arts', the tone of Pattieson's introduction is comic, but it presents a view of artistic and literary production central to the telling of Lucy Ashton's tragic history.[53] By means of the comic exchange between Tinto and Pattieson, Scott puts in motion a series of propositions about the relationship between official history and popular narrative which bear directly on the novel's representation of female insanity.

Dick Tinto's failure to establish himself in the official world of art is comically offset by his success among the artistically uneducated people of the Scottish villages as a painter of pub signs and family portraits. His crude early efforts in this line are a favourite subject of local discussion and regularly occasion a great deal of drunken mirth—so that when, later in life, a more proficient Tinto tries to replace his most embarrassing work, a five-legged horse, his offer is flatly refused by the publican. When Tinto again finds himself 'reduced' to painting pub signs and cheap portraits, Pattieson urges that instead of being ashamed of his character as an 'out-of-doors artist', Tinto should congratulate himself on having brought art to the masses when it too often remains the preserve of the rich. According to Pattieson, the value of art should be calculated by the work's power to generate public interest rather than by its selling-power or by any professional acclaim it may receive. The painter nevertheless flees town rather than stay to witness the hanging of his latest pub sign. Further recollections of Tinto's

[53] Cf. George Levine, *The Realistic Imagination: English Fiction from Frankenstein to Lady Chatterley* (London, 1981), 90.

career also testify to the power of the image not just to compel the attention of the public, but to prompt the production of narrative. Discussing a sketch later recognizable as Ravenswood's confrontation with Lady Ashton when he interrupts the signing of Lucy's marriage contract, Tinto and Pattieson dispute the relationship between a striking artistic impression and the telling of stories. Tinto asserts that every clear image bears within it a clear narrative. Pattieson insists that, on the contrary, a powerful image *instigates* the production of narrative, and has the potential to prompt several conflicting narratives.

The 1830 introduction, added after Scott's public admission of authorship, continues and expands this deceptively casual series of reflections, relating them more directly to Lucy Ashton's story. The introduction pretends to supply information previously withheld about 'the real source from which [Scott] drew the tragic subject of this history': the marriage of Janet Dalrymple to David Dunbar of Baldoon in 1669 and her death less than a month later. Scott gives the following reason for altering his previous determination:

as he finds an account of the circumstances given in the Notes to Law's *Memorials*, by his ingenious friend Charles Kirkpatrick Sharpe, Esq., and also indicated in his reprint of the Rev. Mr Symson's poems, appended to the Description of Galloway, as the original of the Bride of Lammermoor, the author feels himself now at liberty to tell the tale as he had it from connections of his own, who lived very near the period, and were closely related to the family of the Bride. (1)

Scott's references were genuine. He had assisted Sharpe with his editing of Law's *Memorials*, and Sharpe published, in the preface to that volume, several versions of the story upon which *The Bride* was based,[54] but the manner in which the sources are cited is ultimately more important than their content. Though Scott claims to be at last giving 'the real source', he goes on instead to replicate his ingenious friend's efforts by producing not one original story but numerous conflicting ones. His authorities are strangely mixed. He offers references to legitimate historians, all of whom differ in their accounts of the tragedy, but his most compelling and consist-

---

[54] See Coleman O. Parsons, 'The Dalrymple Legend in *The Bride of Lammermoor*', *Review of English Studies*, 73 (1943), 51–8; and James Anderson, *Sir Walter Scott and History, with Other Papers* (Edinburgh, 1981), 70–1.

ent evidence turns out to be common knowledge or hearsay. The established historical records have become, by the end of the introduction, just more voices in a chaos of dubious report and plain misinformation. 'It is well known', Scott begins, that the family of Dalrymple has produced outstanding men in the service of Scotland; he has 'heard the expression used' that the insane bride was 'mopping and mowing'; 'Various reports went abroad on this mysterious affair, many of them very inaccurate, though they could hardly be said to be exaggerated'; 'The credulous Mr Law says'; 'My friend, Mr Sharpe, gives another edition of the tale'; 'I find still another account darkly insinuated in some highly scurrilous and abusive verses, of which I have an original copy.' Scott goes on to quote the 'scurrilous' verses at length, and follows them with yet another example of the questionable literature generated around his subject. And amidst the labyrinth of conflicting testimonies, he keeps generating footnote references to more documentation in his possession.

Neither the reflections on history in the introduction, nor Peter Pattieson's reflections on art and narrative in chapter 1, fit easily with a view of Scott as a rigid conservative in his approach to history.[55] Both present an essentially populist outlook on the production of history, to accommodate which the meanings of images and events are continually reshaped according to the interests of those contemplating them. In line with this emphasis on the popular generation of stories, *The Bride of Lammermoor* is, itself, shaped from the various narratives woven around its central events—many of them highly visual. The motif of portraiture dominates the novel (most obviously in the portrait of Sir Malisius de Ravenswood), as do spectacular phenomena more broadly: ghostly apparitions, disturbing likenesses of the living to the dead. The most powerful moments in the novel have the quality of dramatic paintings, and, as Jerome Mitchell noted, it was these that provided the key to its adaptation for the stage: Lucy sitting by the well like the mermaid of local ledgend; the raven falling from the sky; Edgar bursting in as Lucy signs the last copy of her marriage contract to Bucklaw (the scene Tinto paints). Lucy's madness is the most compelling image of them all, offered as the key to the novel in its introduction, pursued

[55] Cited in Judith Wilt, *Secret Leaves: The Novels of Sir Walter Scott* (London, 1981), 158.

throughout as its *telos*, and continually subjected to conflicting interpretations.

The least developed of those interpretations relates to Lucy herself. Her madness is, of course, at one level, a private emotional affair. Like Marianne Dashwood, Lucy Ashton has a taste for romantic fiction which is both a personal source of strength and a dangerous form of emotional independence, ultimately undermining her, her family, and, indirectly, her society. She finds in her reading a degree of imaginative compensation for her necessary submissiveness to her family:

her passiveness of disposition was by no means owing to an indifferent or unfeeling mind. Left to the impulse of her own taste and feelings, Lucy Ashton was peculiarly accessible to those of a romantic cast. Her secret delight was in the old legendary tales of ardent devotion and unalterable affection, chequered as they so often are with strange adventures and supernatural horrors. This was her favoured fairy realm, and here she erected her aerial palaces. But it was only in secret that she laboured at this delusive, though delightful architecture.... in her exterior relations to things of this world, Lucy willingly received the ruling impulse from those around her.   (40)

Her attraction to Ravenswood is the logical product of that compensatory private taste. But, as Nassau Senior noted, the novel stresses the social context of Lucy's susceptibility to romance greatly at the expense of her claim to be a romantic heroine.[56] So, for example: 'Time, it is true, absence, change of scene and new faces, might probably have destroyed the illusion in her instance as it has done in many others; but her residence remained solitary' (64). Given this reluctance fully to indulge sentimentalism, it might be expected that Lucy's madness would be handled with something of Jane Austen's critical rigour, emerging, like Marianne Dashwood's sickness, as a dangerous, self-destructive, and undesirable form of self-empowerment, albeit the only one available to her.[57] But, however enlightened *The Bride of Lammermoor* may be about the conditions prompting its heroine's taste for romance, her psychology is not the only concern of the novel. Where Austen rebukes the romantic spirit and insists on the

---

[56] See also David Brown, *Walter Scott and the Historical Imagination* (London, 1979), 140–2.

[57] This is the interpretation offered by Philip Martin in *Mad Women in Romantic Writing* (Brighton, 1987), 99–107.

resocialization of her heroine, Scott is far more concerned with exploring the range of other interpretations placed upon her madness and death.

A large proportion of the people of Lammermoor share Lucy's taste for romanticism. Her attachment to a past she fills imaginatively with 'strange adventures and supernatural horrors' mirrors their attachment to superstition and to romanticized accounts of Scotland's past. Interest in romance and superstition appears to be more a feminine characteristic than a masculine one, as the novel presents an ongoing conflict between the men of Wolf's Hope village, whose life-style is becoming more prosperous as they free themselves from ancient feudal dues to the Ravenswood family, and their less enfranchised womenfolk, who cling to romantic tales about Scotland's old established nobility. But the most passionate defender of the Ravenswoods' past is the old family retainer, Caleb Balderstone. The ageing servant exploits the village women's interest in life at the castle to gain access in order to steal a families' celebration dinner for his master's table; when caught, he just saves his skin by playing on their menfolk's vestigial loyalties to the old family.

As Caleb's example might suggest, the urge to cultivate a romantic view of the past and to uphold its claims upon the present is a complex one in *The Bride of Lammermoor*, often most calculated where it seems most pathetically nostalgic. Even the most progressive figures in the novel are repeatedly checked from pursuing their own interests too rigorously by reminders of those they have usurped. Right at the start of *The Bride of Lammermoor*, a chance look at the thirteenth-century portrait of vengeful Sir Malisius causes Sir William Ashton to desist from prosecuting the man's descendants further. History emerges not as the justification of the present system but as its accuser. Recalling the past is primarily the defiant gesture of individuals and groups disinherited by the present order—if, indeed, they ever had access to material power. They summon the past in two ways: either, like Lucy or the gamekeeper, they claim it as a more pleasing age in which they can imagine for themselves a better social condition; or, with greater bitterness, they summon the past to prophesy the downfall of the present social order. Caleb, Alice, Ailsie, the witches, and at times Lucy herself, all use history so. History is the province of the oppressed in this novel, invoked as a mode of opposition to the present order.

It appears a passive mode of opposition, but it bears within it the promptings of a more active revenge. Sir Malisius Ravenswood's words encapsulate the claim to future vengeance that is the most insistent voice of history in this novel: 'I bide my time'.

The cultivation of folkloric history as a source of illicit power underlies the witches' malign delight in Lucy's tragedy. Scott explains the attraction to witchcraft among the poor people of Scotland in terms of their social powerlessness: 'notwithstanding the dreadful punishments inflicted upon the supposed crime of witchcraft, there wanted not those who, steeled by want and bitterness of spirit, were willing to adopt the hateful and dangerous character, for the sake of the influence which its terrors enabled them to exercise in the vicinity, and the wretched emolument which they could extract by the practice of their supposed art' (310). The explanation is not so different from his earlier account of Lucy's attachment to romance. With a far more explicit desire for revenge than Lucy, the witches make use of romanticized accounts of the past in the cause of a quiet class war. Lady Ashton, who alone seems immune to superstition, pays Ailsie Gourlay to use her supposed nefarious influences in breaking down Lucy's fidelity to Ravenswood, and in so doing hastens, if she does not actually cause, her daughter's madness. Gourlay tells the girl the story of the mermaid's well, prophesying disaster for the Ravenswoods, and (Scott suggests more tentatively) shows her Ravenswood's image in a glass, courting another woman. Gourlay and her friends triumph over the Ashtons at Lucy's burial, interpreting her insanity, her murder of Bucklaw, and her death as their own vengeance on the rich: 'Did not I say', says Dame Gourlay, 'that the braw bridal would be followed by as braw a funeral? ... can a' the dainties they could gi'e us be half sae sweet as this hour's vengeance?' (341–2).

Factional violence erupts in the wake of Lucy's attack on her husband, as if her madness unleashes 'the fury of contending passions between the friends of the different parties' that the marriage had promised to subdue (338). Peter Garside has pointed to Scott's interest in the Ashton–Bucklaw marriage as a symbol of Scotland's union with England. As he shows, the language of political debate about the 1707 Act of Union infuses Scott's portrayal of romance and of marriage in *The Bride of Lammermoor*. The possibility of a marriage between Lucy and Ravenswood seems to represent for Scott an ideal of union between England and Scotland, founded in affection and esteem. From the start, however, that hope is threatened by Ravenswood's poverty, and it is quickly ruled out altogether by the new dispensation's contempt for the old Scotland—an attitude exemplified, and vilified, in the figure of Lady Ashton.[58] Caroline Franklin has rightly argued that *The Bride of Lammermoor* also needs to be seen, in the more immediate context of the 1810s, as an expression of its author's pessimism about factional party politics in Scotland during the late 1810s, at a time when unified government seemed more than ever necessary to avert the threat of internal rebellion.[59] Read with these wider political contexts in view, the connections and the differences between Scott's use of the love-mad woman in Scotland, and Maturin and Lamb's earlier use of her in relation to Ireland, become clearer. In *The Bride of Lammermoor*, as in *The Milesian Chief* and *Glenarvon*, rebellion is imaged, through the love-mad woman's story, as, finally, self-destruction. The system survives Lucy's act of madness and her death: Bucklaw recovers from his wounds, vowing never to allow the incident to be spoken of again; and Lady Ashton, who bears the moral responsibility for the tragedy, suffers not at all in her public standing. She has her splendid marble monument to record 'her names, titles and virtues, while her victims remain undistinguished by tomb or epitaph' (349). *Unlike* Maturin and Lamb, however, Scott places a great deal of emphasis on the conflicting stories that will be handed down to the future about the bride of Lammermoor.

There is, of course, one further reading of Lucy's madness—both the most romantic and, finally, the least sustainable within the novel. Edgar Ravenswood's death in the last pages of the novel has very much the quality of an afterword, but it too draws on her insanity for its meaning. Ravenswood spends most of the novel fighting the inclination to join the popular cult of himself as tragically doomed last representative of the old aristocracy but, beaten, finally subscribes to his 'fate'. The cult has an edge of political malice to it, underlined by a series of meetings between Ravenswood and his former tenants. These people (the witches, old

---

[58] See Peter Dignus Garside, 'Union and *The Bride of Lammermoor*', *Studies in Scottish Literature*, 19 (1984), 72–93.
[59] Franklin, 'Feud and Faction'.

Alice, the gravedigger) have suffered far more than he from his family's failure, and that knowledge makes deeply ambivalent their conviction that a malevolent destiny hangs over him. Even Caleb's motivation must be in question. Narrating the Ravenswoods' downfall as a tragedy darkly foreshadowed in the family's history is the one means by which Caleb can restore something of grandeur to Edgar Ravenswood's position and therefore to his own. Hence his quotation of the 'dark words' of verse prophesying Ravenswood's marriage and death:

> When the last Laird of Ravenswood to Ravenswood shall ride,
> And woo a dead maiden to be his bride,
> He shall stable his steed in the Kelpie's flow,
> And his name shall be lost for evermoe!   (185)

The parting between Caleb Balderstone and Edgar Ravenswood underlines the extent to which the nobleman's disinheritance has also been his servant's and, with an ironic twist, the extent to which the servant's refusal to give up the past has made it impossible for Ravenswood to avoid death. Ravenswood's parting words to Balderstone are (with a ghastly smile) 'Caleb! . . . I make you my executor' (347). The double ring of 'executor' is unmistakable.

As the 'dark words' indicate, Lucy has functioned throughout the novel as the sign of Ravenswood's fate, and he and his servant ultimately embrace her insanity and death as their own claim to pathos. In a strong sense, Lucy enables them to follow her to a death which she has ensured will have the status of tragedy.[60] Indeed, their deaths strikingly imitate hers. Ravenswood abandons himself to 'paroxysms of uncontrollable agony' (345), then rides off to encounter his death with the same despairing acquiescence that Lucy took to her wedding. Ravenswood finally has no story apart from that of Lucy—a point underscored by his literal disappearance in the last pages of the novel, absorbed into the quicksands of the Kelpie. Caleb, too, after Ravenswood's death, is a broken man, who spends his days 'moping' about the old castle (the echo of

---

[60] Cf. Alexander Welsh, *The Hero of the Waverley Novels* (New Haven, 1963), 44–5, on Edgar Ravenswood being 'the only hero of the several "tragic" romances who actually shares the unhappy fate of the heroine. Yet the Master of Ravenswood, as he is ironically called, suffers his fate passively, albeit passionately.' Welsh discusses the novel very briefly in a chapter entitled 'The Passive Hero', under the subheading 'Possible Exceptions'.

'mopping' is suggestive), until he 'pines and dies' within a year of the catastrophe.

As the very title of the novel suggests, Lucy Ashton's story belongs to the place in which it is enacted. Her madness and death are less her own desperate rebellion against oppression than the fulfilment of other people's sometimes explicit but as often vague and half-articulated protests against poverty, hardship, and disinheritance. Her madness has finally not one meaning but several. Its very familiarity as a narrative allows it to bring into focus a range of resentments which have no more legitimate means of expression within the official history of Scotland.[61] *The Bride of Lammermoor* thus strongly qualifies any temptation to see Scott as a 'realistic Romantic' who—to quote Georg Lukács's influential assessment— looks soberly at the past and 'finds in English history the consolation that the most violent vicissitudes of class struggle have always calmed down into a glorious middle way'.[62] Rather, *The Bride of Lammermoor* sees romanticism as one means by which both individuals and societies alter the character of their experience, gaining a symbolic though not a material victory over their circumstances. This is a novel fascinated by the underside of history, the 'popular appetite', the desires that must be repressed in any official history that celebrates the 'glorious middle way'. In Scott's words:

By many readers this may be deemed overstrained, romantic, and composed by the wild imagination of an author, desirous of gratifying the popular appetite for the horrible; but those who are read in the private family history of Scotland during the period in which the scene is laid, will readily discover, through the disguise of borrowed names and added incidents, the leading particulars of AN OWER TRUE TALE. (340)

Scott's use of the love-madness convention never fundamentally challenges the accepted grounds for representing female insanity— the woman's loss of her lover; her already romantic disposition—

[61] Cf. John P. Farrell, *Revolution as Tragedy: The Dilemma of the Moderate from Scott to Arnold* (Ithaca, NY, 1980), 114. Farrell argues that in *The Bride of Lammermoor* Scotland cannot revise 'the tragic text of her history' because 'there is no counter-text. The story of community has been edited out.' See also James Kerr, *Fiction against History: Scott as Storyteller* (Cambridge, 1989), 85–6: '*The Bride of Lammermoor* lacks ... any symbol in which the conflicts between classes and political parties can be resolved, or at least held in stable form.'
[62] Georg Lukács, *The Historical Novel*, trans. Hannah and Stanley Mitchell (London, 1962), 32–3.

but it does make an important departure from the way madness has been allowed to figure in fiction for the previous half-century or more. In a very different way from Jane Austen, he makes his novel acknowledge an element of social determination about female insanity. In *The Bride of Lammermoor*, Lucy's madness is the most fitting expression available for a distress that has otherwise no social visibility and no official history; and the novel is acutely aware of the politically dangerous sympathies that the familiar outline of her story might be capable of conveying. Scott presents the love-mad woman as a powerful symbol of the breakdown in relations between men and women, between political factions, and between classes, that threatens an ostensibly progressive society. Different in kind though they are, her distress makes possible the narration of a wider social distress, so that this 'private family history of Scotland' becomes, more accurately, the unofficial public history of Scotland.

5

The Hyena's Laughter: Lucretia *and* Jane Eyre

THE explicit association of love-mad women with insurrection slowly faded during the 1820s. Stories about them continued to attract writers, and although, in the absence of a comprehensive survey of 1820s and 1830s fiction, it is difficult to be confident about the numbers of novels concerned, a general pattern emerges. Walter Scott's influence was strong. The madwomen in Bulwer-Lytton's *Godolphin* (1833); Rosina Lytton's *Cheveley; or, The Man of Honour* (1839) and Ellen Pickering's *Nan Darrell; or, The Gipsy Mother* (1839) are discernibly indebted to Madge Wildfire and Meg Merrilies. Lady Caroline Lamb also had her followers, notably W. H. Maxwell, whose *O'Hara; or, 1798*, published anonymously in 1825, drew heavily on *Glenarvon*.[1] As these examples show madwomen were moving down the social hierarchy in British fiction of the 1820s and 1830s, often becoming either gypsies themselves (Scott's Meg Merrilies,[2] Pickering's Nan Darrell) or closely associated with gypsies (Lytton's Mary Lee).[3] They were rarely now the central figures of new fiction, and, interestingly, they seem to have been ageing. *O'Hara, Nan Darrell*, and Frances Trollope's *Michael Armstrong* (1839) all depict crazy old women, whose age appears to give literal expression to their long history in English fiction. More importantly, in terms of where the convention would go next, many of these women act out revenge plots from which the novels' more aristocratic heroines are carefully distanced.

---

[1] An interesting exception is Frances Trollope's industrial novel *The Life and Adventures of Michael Armstrong, The Factory Boy* (1839), which made effective use of pathetic female insanity as one weapon in a searing attack on conditions in the industrial North. Old Sally's insanity arises not from maltreatment by a lover, but from her exploitation since childhood by the owners of the textile mills.

[2] From *Guy Mannering; or, The Astrologer* (Edinburgh, 1815).

[3] The figure of the love-mad gypsy persisted in English fiction. For a late example, see James Payn's *Lost Sir Massingberd: A Romance of Real Life* (London, 1864).

Already evident in *Glenarvon* and *Madelina*, the pairing of a sane heroine against a deranged other woman had become something of a standard feature in popular romance by the time the most famous of all novels about female insanity appeared in 1847.

Critical writing about madwomen in the nineteenth-century novel has latterly been dominated by the example of *Jane Eyre*— usually with little or no sense that Charlotte Brontë might have been writing on a subject which had attracted numerous writers of fiction, poetry, and drama before her. When Nancy Armstrong claims in *Desire and Domestic Fiction* (1987) that 'deranged women suddenly came into vogue with the great domestic novels of the 1840s', she is clearly unaware of the mass of earlier material on the subject.[4] Armstrong is not alone in finding *Jane Eyre*'s madwoman a difficult figure to place. For most literary critics, Bertha Mason is a strange incorporation of gothic theatricality into domestic fiction—a figure from the past who, in her very strangeness, can stand as an exemplary expression of the rebellion and rage seething within the woman writer's unconscious,[5] or an equally ahistorical exemplum of 'mythic' representations of femininity.[6] In this context Armstrong's book stands out as a rare attempt to account historically for the violence attaching to mid-nineteenth-century representations of female insanity.[7]

Desire and Domestic Fiction sees the monstrous female maniac of mid-Victorian fiction as the indirect product of a long-standing English rural tradition in which political grievances were dramatized through theatrical displays of sexual anarchy—a tradition which came under increasingly hostile government scrutiny after 1789. Armstrong argues that, in the wake of the Peterloo Massacre of 1819, sociologists and other writers sought to promote a more

---

[4] Nancy Armstrong, *Desire and Domestic Fiction: A Political History of the Novel* (Oxford, 1987), 164. Armstrong goes on to discuss *Oliver Twist*, *Wuthering Heights*, and *Mary Barton*, none of which can strictly be said to depict female insanity.

[5] Most influentially, Sandra M. Gilbert and Susan Gubar, *The Madwoman in the Attic: The Woman Writer and the Nineteenth-Century Literary Imagination* (New Haven, 1979).

[6] Notably Nina Auerbach, *Woman and the Demon: The Life of a Victorian Myth* (Cambridge, Mass., 1976), and Bram Dijkstra, *Idols of Perversity: Fantasies of Feminine Evil in* Fin-de-siècle *Culture* (Oxford, 1986).

[7] For an earlier and in some ways similar example, see also Eric Trudgill, *Madonnas and Magdalens: The Origins and Development of Victorian Sexual Attitudes* (London, 1976).

benign concept of power, one less centralized, less repressive, and less dogmatic.[8] Their answer, she claims, was to lay heavy stress on domestic harmony as a model for industrial, social, and political behaviour, with the result that the fear of political insurrection was more often represented not in direct terms but in coded form, translated into images of sexual scandal. The spectacle of demonic female madness is Armstrong's primary example of that translation at work. The monstrous madwoman emerges as a creation of domestic fiction of the 1840s, where she expresses a capacity for violence that has been expelled as far as possible from the political arena. She is, in Freudian terms, the return of the repressed in English culture of the mid-nineteenth century.[9]

Armstrong's genealogy of the 1840s madwoman intersects with an expansive literature on Victorian domesticity, much of it focused on the peculiar dualism at work in the period's representations of femininity.[10] Most criticism to date has made little or no attempt to explain the Victorians' fascination with images of demonic femininity, and in this respect Armstrong's account of Bertha Mason is, again, a welcome exception. As a historical explanation, however, it is seriously flawed. In the first place, there was no simple displacement of 'politics' into 'culture' after 1819. English radicalism did not disappear during the 1820s and 1830s. Rather, these were years of growing social distress and economic upheaval, and working-class agitation was a continual fact of British politics. Chartism came to dominate British political consciousness, not only during 1838–9 when agitation was at its strongest, but long after, and those who were anxious about radicalism had plenty to fuel their alarm in newspaper and journal accounts of revolution across Europe in 1830, and again in 1848. There was, in other words, no sudden 'domestication' of politics in early Victorian England that could explain the emergence of a newly violent conception of female insanity in fiction.

[8] Armstrong, *Desire and Domestic Fiction*, 176.
[9] This argument has proved formative, surfacing most recently in Linda Shires's essay on Victorian representations of the French Revolution, in Linda Shires (ed.), *Rewriting the Victorians: Theory, History, and the Politics of Gender* (New York, 1992), 147–65.
[10] Amanda Vickery, 'Golden Age to Separate Spheres? A Review of the Categories and Chronology of English Women's History', *Historical Journal*, 36 (1993), 383–414, provides an excellent survey and bibliography of the vast literature on this subject.

It is equally important to question the scale of the 'vogue' for Bertha Mason-like madwomen in novels of this period. Reading any of the numerous critics who have seen in the 1840s the germs of the *fin de siècle*'s obsession with demonic women, it would be easy to assume that brutal and violent madwomen were two a penny. In fact, very few novelists produced anything comparable to Charlotte Brontë's description of female insanity. The chronically impoverished 'queen' of cheap fiction, Hannah Maria Jones, spotted the potential market and tried to tap Brontë's success with *The Trials of Love; or, Woman Rewarded* (1849), but even she used her bloated, gin-mad Irishwoman, Rose Sinclair, for a *tour de force* ending and chose not to allow her a more expansive role in the narrative. The only work which genuinely bears close comparison with *Jane Eyre* in its depiction of female insanity is Edward Bulwer-Lytton's *Lucretia; or, The Children of Night* (1846). Published a year before *Jane Eyre*, *Lucretia* strikingly anticipates its more famous successor.

The fact that these two works found so few imitators may indicate the degree to which their depictions of female insanity were shaped by conflicting and difficult pressures. Female insanity was not an easy subject for fiction in the 1840s. In so far as *Lucretia* and *Jane Eyre* invoke the sentimental and now *déclassé* figure of the love-mad woman, they do so nostalgically or, more often, in dark parody. Both works are nevertheless heavily influenced by fiction of the 1810s and 1820s. The extent of this debt has been virtually ignored in recent criticism, yet the issues that had preoccupied Sir Walter Scott, Charles Maturin, and Lady Caroline Lamb are deeply embedded in the language, the imagery, the symptoms, and the associations of female insanity in *Lucretia* and *Jane Eyre*. For their first readers, the association between female madness and murderous violence may have been shocking but it would not have been new. Elizabeth Rigby's much-quoted analogy between Chartism and 'the tone of the mind and thought' at work in Brontë's novel[11] would have seemed to many of her contemporaries not forced but only too apparent.

Edward Bulwer-Lytton's *Lucretia; or, The Children of Night* had a harsh reception when it first appeared. Bulwer-Lytton had an estab-

---

[11] Elizabeth Rigby, unsigned review in the *Quarterly Review*, 84 (Dec. 1848), 153–85; repr. in Miriam Allott (ed.), *The Brontës: The Critical Heritage* (London, 1974), 109–10.

lished reputation as the author of several best sellers of the 1820s and 1830s: a mixed bag which included the hugely popular 'silver-fork' novel *Pelham* (1828), the scandalous and even more successful Newgate novels *Paul Clifford* (1830) and *Eugene Aram* (1832), and the historical novel *The Last Days of Pompeii* (1834). He had acquired some fashionable notoriety as an ex-lover of Lady Caroline Lamb, and some less welcome publicity in the late 1830s from the acrimonious breakdown of his marriage to Rosina Wheeler, a protégée of Lamb. In the mid-1840s he profited from a break in his parliamentary career by pouring out a stream of plays, mystical novels, and journalism; but with *Lucretia* he returned to the genre which had brought him such success in the early 1830s, the Newgate novel.

The *Athenaeum* summed up the general consensus when it pronounced *Lucretia* 'a bad book of a bad school', and one which would imperil the author's reputation.[12] Bulwer-Lytton replied with a pamphlet entitled *A Word to the Public*, attempting to justify the work as a deliberate departure from realism, unjustly censored according to realist criteria. He was answered in turn by a leading article in the *Westminster and Foreign Quarterly Review*, which asserted the moral obligations of all art, and particularly of tragedy.[13] Edwin Eigner notes that several letters from John Forster at this time suggest that Bulwer-Lytton was so crushed he believed he would never write again. When he did so it was to complete a book begun at the same time as *Lucretia* but firmly within the genre of domestic fiction. *The Caxtons* succeeded in rescuing his damaged reputation.[14] He completed his penance by revising *Lucretia* in

[12] Unsigned review in the *Athenaeum*, 5 Dec. 1846, 1240–2. *Lucretia* did have a few admirers. Mrs Henry Wood and Mrs Maxwell approved of it as a real 'blood-curdler'; quoted in T. H. S. Escott, *Edward Bulwer First Baron Lytton of Knebworth: A Social, Personal, and Political Monograph* (London, 1910), 5. Macaulay wrote to Bulwer-Lytton saying he would place *Lucretia* 'very high' among the author's achievements: 'it is some years since any fiction has made me so sad'; letter to Bulwer-Lytton, 14 Dec. 1846, quoted in Victor Alexander, 2nd Earl of Lytton, *The Life of Edward Bulwer, First Lord Lytton, by his Grandson the Earl of Lytton* (London, 1913), ii. 92–3: 93.

[13] Unsigned article, 'The Province of Tragedy: Bulwer and Dickens', *Westminster and Foreign Quarterly Review*, 47 (Apr. 1847), 1–11: 2.

[14] Edwin M. Eigner, 'Bulwer's Accommodation to the Realists', in Harold Orel and George W. Worth (eds.), *The Nineteenth-Century Writer and his Audience* (Lawrence, Kan., 1969), 66.

1853, reducing its death-toll in a gesture of goodwill towards his outraged public.

Lucretia is the grim story of a woman's downward spiral from immorality into insanity. Lucretia Clavering, a clever but corrupt English girl, is disinherited by her uncle when he is alerted to her unfilial behaviour. At the same time she is rejected by the man she loves. The dual rejection scars her, emotionally and morally, for life. On the rebound, Lucretia marries her tutor, Oliver Dalibard, an ex-crony of Robespierre escaped from France, and a man of extraordinary mental powers and vicious character. He educates her in crime but gradually earns her bitter hatred. She murders Dalibard, remarries, and poisons her second husband. Before he dies, this man manages to remove their child secretly to a safe house, and all Lucretia's efforts to locate the boy are unavailing. She is found, almost mad with maternal grief, by Dalibard's son, Gabriel Varney, and together these two establish a highly successful partnership of crime. The threads of Lucretia's story are next picked up several years later. Posing as a cripple, she attempts to murder the daughter of the man she once loved and has never forgiven. Her plans are thwarted by a crossing-sweeper named Beck—a clear precursor of Jo in Dickens's *Bleak House*. In revenge, Lucretia stabs Beck with a poisoned ring, only to discover, too late, that he was her lost son. She promptly becomes a raving maniac, and the novel ends with a lurid description of her future as the 'cureless and dangerous' inmate of a madhouse. Gabriel Varney escapes to the New World where he is tortured to death by savages.

The models for Gabriel Varney and Lucretia Dalibard have been identified as Thomas Wainewright (1794–c.1849) and his wife Frances. Wainewright first came to public attention in the early 1820s when he exhibited pictures at the Royal Academy and contributed art criticism to the *London Magazine*. He cultivated the 'dandy' and 'silver-fork' styles of writing—a taste which, as Hazlitt noted, gave cause for comparison with Bulwer-Lytton.[15] He was a friend and patron of Blake, and he mixed with Lamb, Hazlitt, Macready, and De Quincey. In 1837, however, he was convicted of forgery and transported to Australia for life. Wainewright may well

---

[15] W. Carew Hazlitt (ed.), *Essays and Criticisms by Thomas Griffiths Wainewright* (London, 1880), p. xxxvi.

or a subject which would adequately represent 'that crisis
to which we have arrived', he 'became acquainted with
es of two criminals' (vol. i, pp. vii–viii): 'it seemed to me,
ng their lives, and pondering over their own letters, that
heir cultivation itself we could arrive at the secret of the
nd atrocious pre-eminence in evil these Children of Night
ed—that here the monster vanished into the mortal, and
omena that seemed aberrations from nature were ex-
vol. i, p. ix). The Carlylean rhetoric[21] failed to impress
ytton's critics, and it is certainly not borne out by what
although the first chapters are clearly concerned to depict
Clavering and Oliver Dalibard as extreme pathological
of capitalist greed, the novel never resolves the tension
ere between 'the monstrous' and 'the mortal', 'the evil'
erely 'criminal'. In Lucretia's case, particularly, the move
rehend the monstrous in human terms proves un-
le. Bulwer-Lytton tampered extensively with his source
n order to make the woman, rather than her husband, the
igure of interest. A murderess evidently seemed to him a
matically striking figure than a murderer, and more likely
he kind of answers he was looking for, but in practice
proved a far less manageable subject than her male
ons.

a's story belongs in its outline to the narrative conventions
ve-mad woman and the mad mother, but she is barely
d by them. Bulwer-Lytton's heroine presents a continual
in terms of sexual representation. The novel gives her a
for criminal cunning which it denounces as masculine and,
y, monstrous,[22] yet it insists on making a bid for sympathy
ng sentimentally to the woman she would have been if not
ss of her first lover and the corrupting tutorship of Oliver
. So, Lucretia has 'none of the sweet feminine habits which
lovelily the whereabout of women' (ii. 26).[23] Allusions to
Borgia, Lady Macbeth, and Clytemnestra are recurrent,[24]
prepared to condemn Lucretia utterly, the novel draws a

are also clear similarities with Disraeli's writing. *Lucretia* coined the
es and have-nots'.
id. i. 103–6 and 227; iii. 90.
ibid. iii. 98.
id. ii. 206 and 209–18.

have been relieved at his sentence, given that he had escaped pros-
ecution for the poisoning of his 21-year-old half-sister, Helen
Abercrombie, supposedly for insurance gain (he was also suspected
of having murdered his mother-in-law and an uncle). While awaiting deportation in Newgate, Wainewright was recognized by
Macready, who was being shown over the prison with Charles
Dickens and John Forster. Rumour soon had it that on this occasion Wainewright effectively admitted poisoning Helen, and
'urged in extenuation that she had very thick ankles'.[16] As a
friend of all three of the Newgate visitors, Bulwer-Lytton no doubt
heard the story. His research for his novel also involved a correspondence with one of the insurance companies involved, and
in 1849 he wrote to inform the office that he had just learned
of Wainewright's death in an Australian hospital: 'His latter days in
the sick ward were employed, I am told, in blaspheming to the
pious patients and in terrifying the timid. I think that he never lived
to know the everlasting fame to which he has been damned in
Lucretia.'[17]

Lucretia was therefore highly topical, but it also signals in its
prologue a continuing interest in the political anxieties which had
shaped so much fiction over the previous few decades, and a concern to set the immorality of the present day in that larger context.
In Paris at the height of the Reign of Terror,[18] Oliver Dalibard takes
his small son to witness the execution of a woman whom the child
only gradually recognizes as his mother. The pretty, frail English
dancer has been unfaithful to Dalibard, conducting a secret liaison
with a 'çi-devant marquis' whom she has concealed from the police.
On discovering her infidelity, Dalibard has betrayed them both to
the government, and brought his son to witness the double execution. 'I must get you a good place for the show,' he tells Gabriel,
and the ensuing description stresses this production of the woman's
death and her terrified madness as melodrama:

then the crowd caught sight of [the marquis's] companion, who was being
lifted up from the bottom of the tumbril, where she had flung herself in

[16] Sidney Lee (ed.), *Dictionary of National Biography*, 58 (London, 1899), 437–
9: 439.
[17] 2nd Earl of Lytton, *The Life of Edward Bulwer, First Lord Lytton*, ii. 88.
Bulwer-Lytton's letter throws doubt on the date of Wainewright's death as given in
the *Dictionary of National Biography*.
[18] On Bulwer-Lytton's interest in the French Revolution, see Allan Christensen,
Edward Bulwer-Lytton: The Fiction of New Regions (Athens, Ga., 1976), 114–15.

horror and despair. The crowd grew still in a moment, as the pale face of one, familiar to most of them, turned wildly from place to place in the dreadful scene, vainly and madly through its silence, imploring life and pity. How often had the sight of that face, not then pale and haggard, but wreathed with rosy smiles, sufficed to draw down the applause of the crowded theatre—how, then, had those breasts, now fevered by the thirst of blood, held hearts spellbound by the airy movements of that exquisite form writhing now in no stage-mime agony! . . . Butterfly of the summer, why should a nation rise to break *thee* upon the wheel?   (i. 13)

The woman's forced exchange of a playhouse stage for an executioner's block reproduces a long-standing association between the theatre and the scaffold. The rapt attention with which she was met as an actress seems also to apply here: those who crowded the theatre in the past now crowd the revolutionary square; they watch with equal attention now as then 'that exquisite form writhing', their involvement in the scene (breasts fevered by the thirst not '*for* blood' but '*of* blood') suggesting sexual desire as much as the desire to witness a death. Bulwer-Lytton's writing exploits somewhat tastelessly the shock of the difference between what this woman so recently was and what she is now: her ability to 'draw down' the applause of the mob was formerly seductive, now her situation seems closer to rape; her performance expresses 'no stage-mime agony' but genuine terror; her face, once 'wreathed with rosy smiles', is 'pale and haggard'.

    Thus far, the prologue is distinctly Burkean in its disparagement of the French revolutionary mob. The allusion to Pope's satire on Lord Hervey ('Satire or Sense, alas! can *Sporus* feel? | Who breaks a Butterfly upon a Wheel?'[19]) underscores a critical detachment from the scene which seems initially to belong only to the narrator and reader. But Bulwer-Lytton's crowd gradually begins to acknowledge a 'sense of the mockery of such an execution, of the horrible burlesque that would sacrifice to the necessities of a mighty people so slight an offering' (i. 13–14). A sentimental conviction that wreaths of smiles are the only appropriate wreaths for this 'butterfly of the summer' soon undermines the belief that political justice is being done. To conceive of the airy dancer as political traitor is unpalatable even to this bloodthirsty crowd, and, with the woman's dismissal as a political insurgent, what has been a public

[19] 'An Epistle to Dr Arbuthnot', in John Butt (ed.), *The Poems of Alexander Pope*, iv: *Imitations of Horace*, 2nd edn. (London, 1961), 118.

---

spectacle of revenge abruptly shi[...] one:

The dangerous sympathy of the m[...] attendance. Hastily he made the sign [...] child's cry was heard in the Englis[...] Father's hand grasped the child's arn[...] swam before the boy's eyes; the air see[...] red; only through the hum, and the t[...] heard a low hiss in his ear—'Learn h[...]

    As the father said these words, agai[...] whose ear, amidst the dull insanity of f[...] voice, saw that face, and fell bac[...] headsman.   (i. 14)

The woman's political crime agai[...] child's cry, a private, emotional iss[...] index of the shift is an English chil[...] of the French mob, for the appeal t[...] domestic values in this novel. The n[...] signal from the hostile perspectiv[...] subjective—indeed, empathetic—ga[...] mother's, swim; he, too, feels 'an i[...] as it must her, and grows blood-red.[...] vengeance at the end seems aimed [...] woman, and the rest of the novel be[...] psychological perversion of the chil[...] Gabriel will prove to be a more vici[...] than his father. It also bears testimo[...] perversion and criminality as equiva[...] feminine.[20]

    The prologue's shift in emphasis fr[...] a more private revenge drama matche[...] ing sense of what the novel's focus sh[...] preface to the first edition of *Lucretia*[...] been to trace 'the strange and secret w[...] ruler of Civilization, familiarly callec[...] into our thoughts and motives, our he[...]

[20] Gabriel Varney grows to be a more dange[...] Dalibard's ruthless ambition with an 'almost [...] him to manipulate women more effectively an[...] an implication of homosexuality.

Looking [...]
of societ[...]
the histo[...]
on study[...]
through [...]
ruthless [...]
had atta[...]
the phe[...]
plained'[...]
Bulwer-I[...]
follows.[...]
Lucretia[...]
products[...]
evident [...]
and the [...]
to comr[...]
sustaina[...]
material[...]
primary[...]
more dr[...]
to yield[...]
Lucretia[...]
compan[...]

    Lucre[...]
of the l[...]
containe[...]
problem[...]
capacity[...]
ultimate[...]
by allud[...]
for the [...]
Dalibar[...]
betray s[...]
Lucreti[...]
but, no[...]

[21] The [...] terms 'ha[...]
[22] e.g. [...]
[23] Alsc[...]
[24] e.g. [...]

strategic distinction between mind and body: Lucretia is 'more intellectually unsparing than constitutionally cruel, (save where the old vindictive memories thoroughly unsexed her)' (ii. 274).[25] The struggle between womanliness and evil is continual (she shudders with horror when she contemplates crime; ii. 24), and the resulting instability of tone repeatedly undercuts Bulwer-Lytton's writing. As the novel progresses, the narrative lurches between subjective, sympathetic rendering of her emotional conflicts, and distanced condemnation of an irredeemable criminality.[26]

Mental vulnerability finally prevents Lucretia Clavering becoming another Lucretia Borgia. Her long history of crime ends in madness with the discovery that Beck was her son:

with a low suffocated cry, [the boy] slid from the hand of Ardworth, and, tottering a step or so, the blood gushed from his mouth, over Lucretia's robe;—his head drooped in an instant, and falling, rested first upon her lap—then struck heavily upon the floor. The two men bent over him, and raised him in their arms—his eyes opened and closed—his throat rattled, and, as he fell back into their arms a corpse, a laugh rose close at hand—it rang through the walls, it was heard near and afar—above and below. Not an ear in that house that heard it not. In that laugh fled for ever, till the Judgment-day, from the blackened ruins of her lost soul, the reason of the murderess-mother.   (iii. 272)

Lucretia (as Bulwer-Lytton crossly insisted) is not a realistic novel, but as this passage demonstrates it *is* intimately concerned with domestic values. More precisely, it is about the definition of domesticity *as* realism. The last phrase sums up the conflicts of representation that have arisen around its attempt to envisage a self-possessed criminal woman. Though the preface claims that 'the monster vanishes into the mortal' the logic of Bulwer-Lytton's novel is rather the reverse. Flamboyantly overwritten, few passages of nineteenth-century fiction more openly demonstrate the gesture on which the Victorian cult of domestic virtue depended: the expulsion of female criminality into the realm of monstrosity. A Medea without tragic scope or stature, Lucretia is thoroughly denatured in a description of insanity which closely anticipates the pealing laughter and the blackened ruins that will accompany Bertha Mason's insanity.

---

[25] See also ibid. ii. 164.   [26] e.g. ibid. ii. 24.

When the last chapter describes Lucretia in the asylum, the suddenly very obtrusive narrator alerts the reader to the novel's departure from established literary guide-lines, underlining the degree to which its violence exceeds the most grotesque models available from the past. The visitor-reader is guided through the asylum, stopping to 'sigh' nostalgically before the familiar gentle maiden of sentimental fiction who smiles insanely at a delusive vision of her shipwrecked lover's return. Then

—you pass by strong grates into corridors gloomier and more remote. Nearer and nearer, you hear the yell, and the oath and blaspheming curse—you are in the heart of the Mad-house, where they chain those at once cureless and dangerous—who have but sense enough left them to smite, and to throttle, and to murder. Your guide opens that door, massive as a wall, you see (as we, who narrate, have seen her) Lucretia Dalibard:— a grisly, squalid, ferocious mockery of a human being—more appalling and more fallen, than Dante ever fabled in his spectres, than Swift ever scoffed in his Ya-hoos!—   (iii. 298)

'Cureless and dangerous', Lucretia in her madness cannot be described other than by her literary excess, 'more appalling and more fallen' even than the fabulous spectres of Dante or the Yahoos of *Gulliver's Travels*. Even fantasy fails to supply a vocabulary for this 'ferocious mockery' of humanity. At the crux of the asylum scene, the narrative voice interposes for the first time a claim to the authority of more than one pair of eyes: 'we, who narrate, have seen her'. But the mad eye of Lucretia overpowers all others:

Only where all other feature seems to have lost its stamp of humanity, still burns with unquenchable fever—the red devouring eye. That eye never seems to sleep, or, in sleep, the lid never closes over it. As you shrink from its light, it seems to you as if the mind that had lost coherence and harmony, still retained latent and incommunicable *consciousness* as its curse. For days, for weeks—that awful maniac will preserve obstinate, unbroken silence; but . . . sometimes [her hands] gather up the hem of that sordid robe, and seem, for hours together, striving to rub from it a soil. Then, out from prolonged silence, without cause or warning, will ring, peal after peal (till the frame, exhausted with the effort, sinks senseless into stupor), the frightful laugh. . . . Baffler of man's law, *thou*, too, hast escaped with life! Not for thee is the sentence, 'Blood for blood!' Thou livest—thou mayst pass the extremest boundaries of age. Live on, to wipe the kiss from thy brow, and the blood from thy robe!—LIVE ON!   (iii. 298–9, 302–3)

As Lucretia's eye mutates into the evil eye, the novel finally condemns its own creation, the heavy archaisms and the capitalization of 'Live on' underlining its curse on this child of night. Even here, however, the codes of representation are oddly incoherent. The grotesque and the fantastic mix strangely with remnants of the pathetic, so that Lucretia still retains the feminine gesture of striving to clean the hem of her robe, when every other sign of femininity is gone (the allusion to Lady Macbeth again reinforces the transgression of her sex). Lucretia does, as the text acknowledges—and as its beleaguered syntax confirms—baffle man's law.

With its uneasy handling of Lucretia Dalibard, Bulwer-Lytton's novel registers an important change in the cultural significance of its subject. The problems the madwoman presents in this novel are, above all, evidence of the conflicting messages she brings to the late 1840s. Lucretia's story repeatedly exploits romantic precedents in which the abandoned woman finds an alternative outlet for her disordered emotions in subverting the state. Bulwer-Lytton was clearly divided over the justice of any claim she might have to sympathy. For most of the novel, he does his best to deprive her of sentimental attraction; yet there are also signs that he wanted to hold on to the older, more sympathetic image of love-madness (the girl in the asylum, pathetically smiling) which would enable him to humanize his subject. It is tempting to see *Lucretia* simply as being poised between two models: one in which a woman's madness is a culturally persuasive metaphor for a diseased body politic, corrupt and capable of revolutionary ferment; and a more familiarly Victorian one in which her madness expresses a private, at most familial, tragedy. In fact, *Lucretia* lays bare the impossibility of that erasure of history. If this woman's criminality is a sign of the times ('that crisis of society to which we have arrived') it is because she is among the most potent symbols available to express the continuing fear of political insurgency that unsettles any attempt to construe her actions as 'merely' familial and private.

The 'merely private' has a way of coming back when it is least wanted. Ironically, many of *Lucretia*'s first readers would have found it difficult not to interpret the novel as a response to the author's own domestic misfortunes. Edward Bulwer-Lytton is known to have been familiar with earlier nineteenth-century fiction about female insanity. He wrote enthusiastically of Scott's *The*

Bride of Lammermoor;[27] and he recorded his impressions of *Glenarvon* on the flyleaf of his own copy. He noted that when he first read it at school, it 'made on me a deeper impression than any romance I remember, and, had its literary execution equalled the intense imagination which conceived it—I believe that it would have ranked amongst the few fictions which produce a permanent effect upon youth in every period of the world'.[28] But there were more personal reasons for his associating the convention of the love-mad woman with the public expression of a private unhappiness—and in order to understand *Lucretia* fully it is necessary to refer back to an equally obscure novel of the late 1830s. When relations between Lord and Lady Lytton broke down in 1836, Rosina's attacks on her husband included the penning of several vicious *romans-à-clef*. The most notorious of her literary efforts was *Cheveley; or, The Man of Honour* (1839). Thackeray read the manuscript for the publisher, James Fraser, and strongly deplored the idea of publication, although he was no admirer of Bulwer-Lytton.[29] Undeterred, Lady Lytton published *Cheveley* at breakneck speed with Edward Bull in London and with Baudry's European Library in Paris (she got her own back on Thackeray by lampooning him in the novel as 'lick-dust' Fuzboz). The three volumes totalling around 1,000 pages were completed in nine days 'although every proof-sheet had to be forwarded to the author for correction, and to travel, to and fro, the distance of two hundred and sixteen miles'.[30] *Cheveley* proved the *succès de scandale* Rosina had hoped for, going through three editions within six months. Prefaced by a scalding attack on the male sex (addressed 'To No One Nobody, Esq., of No Hall, Nowhere'; p. v) the novel is, from

[27] On Bulwer-Lytton's reading of *The Bride of Lammermoor*, see his letter to his son of 13 Feb. 1866, in 2nd Earl of Lytton, *The Life of Edward Bulwer*, ii. 367; and on *Glenarvon*, i. 120.

[28] Bulwer-Lytton's copy of *Glenarvon* is held in the library of Knebworth House.

[29] His reading of the MS is recorded in an unpublished letter sold at auction at Quaritch's in July 1960 (*Quaritch's English Literature and History*, catalogue no. 807, p. 31, item 411). Thackeray was no admirer of Bulwer-Lytton. He attacked him several times, most famously in *Punch's Prize Novelists*, where he parodied him as 'Sir E. L. B. L. BB. LL. BBB. LLL., Bart.'; *Punch*, 3 Apr. 1847, repr. in *The Oxford Thackeray*, viii: *Miscellaneous Contributions to Punch 1843–1854*, ed. George Saintsbury (London, 1908), 83. Thackeray had, nevertheless, always had a sneaking fascination with the object of his satire, and he used the preface of his 1852 *Complete Works* to apologize.

[30] Edward Bull, *Hints and Directions for Authors* (1842), quoted in Allan C. Dooley, *Author and Printer in Victorian England* (Charlottesville, Va., 1992), 17.

first to last, an attempt to pillory Bulwer-Lytton, and it takes as its weapon the convention of the love-mad woman.

In the guise of Lord De Clifford, Lord Lytton storms, seduces, and brutalizes his way through *Cheveley*, a tyrant to everyone about him, but above all to his beautiful wife Julia (Rosina). De Clifford's downfall is his villainous seduction and betrayal of Mary Lee, the daughter of one of his tenants. Disguised as a Norfolk farmer, he woos her, and goes through a mock ceremony of marriage with her. When she finds herself with child, he writes a letter brutally renouncing all responsibility and advising that she take herself off to a House of Correction. Mary Lee at first declines into an old-style sentimental love-melancholy, but when she discovers the true identity of her seducer, she eschews pathos in favour of revenge (Rosina Lytton had also read her Lamb and her Scott). Assisted by a thoroughly Scottian gipsy named Madge, Mary dabbles in witchcraft and conjures up the prophetic phantasmagoria of Lord De Clifford falling from his horse and streaming with blood. Sure enough, several years later, vengeance has its day, and the madwoman denounces her oppressor in front of a full courtroom. He flees the room, leaps on to his horse, and falls to his death a few hundred yards down the road. In another foreshadowing of *Jane Eyre*, De Clifford's corpse resembles the damaged body of Mr Rochester when Brontë's madwoman has exacted her revenge: 'the forehead and one eye were completely smashed into the head' (*Cheveley*, 322).

Bulwer-Lytton may well have been responsible for an anonymous poem published hot on the heels of *Cheveley*.[31] *Lady Cheveley; or, The Woman of 'Honour'* heaped opprobrium on Lady Lytton for her scandalously unwifely behaviour, and thundered to the defence of her Lord. Among its specific targets was 'the over-drawn and improbable story of Mary Lee, which it is needless to contradict, since it so often contradicts itself'.[32] Perhaps, but the poem nevertheless took some pains to contradict the story:

> Yes, when his warm and gen'rous hand supplied
> The wedding portion of a peasant-bride,
> The deed so worthy of his noble name

[31] See Marie Mulvey Roberts, Introduction to Lady Lytton Bulwer (Rosina Lytton), *A Blighted Life: A True Story* (Bristol, 1994), p. xxiii.

[32] *Lady Cheveley; or, The Woman of 'Honour'. A New Version of Cheveley, the Man of Honour* (London, 1839), 45 n.

> Was made the base of such a tale of shame;—
> Like charnel lights, that from corruption shine,
> It glares from Cheveley's page—'twould sully mine![33]

If Bulwer-Lytton was the author, he was clearly rattled. But, as will become apparent in Chapter 6, this was not the last, or the most serious, instance in which female insanity would become a weapon in the Lyttons' vicious domestic squabbles, and Bulwer-Lytton would have far more reason to fear for his good name before it was all over.

Lucretia could not directly have influenced *Jane Eyre*—and it is highly unlikely that Charlotte Brontë would have tolerated *Cheveley* for more than two pages. *Jane Eyre* was already in the hands of its publisher, Smith, Elder, when Bulwer-Lytton's novel appeared. Several early reviewers were quick to point out the differences between the two writers,[34] but there were also similarities. Quite apart from the remarkable parallels between their plots, both writers were (surprisingly) often categorized as authors of 'domestic fiction', yet both caused considerable concern among reviewers by their failure to exhibit the moral standards expected of that genre.[35] The shifting cultural significance of female insanity, and Bulwer-Lytton's own uncertainty about the degree of sympathy that could be afforded to Lucretia, seriously undermine his novel's coherence. In *Jane Eyre*, the instability of the madwoman's meaning for mid-nineteenth-century English fiction is more firmly held in view and turned to effect. Brontë was much the younger writer. She was 30 years old in 1846, whereas Bulwer-Lytton was 43, and although he was critical of the inflated sub-Byronic style of *Glenarvon*,[36] his own writing suffers from the same hyperbolic excesses. Brontë, by comparison, grew up during the last years of high Romanticism (Byron died two days before her eighth birthday), and, for her, growing up involved the gradual cultivation of rather more scepticism towards the literature that had so much absorbed her during childhood and adolescence.

[33] *Lady Cheveley*, 45.
[34] See the unsigned review in *Era*, 14 Nov. 1847, 9, and the unsigned review, probably by A. W. Fonblanq, in the *Examiner*, 27 Nov. 1847, 756–7; repr. in Allott, *The Brontës*, 77, 79. For further examples, see pp. 82, 213–14, and 444.
[35] Allott, *The Brontës*, 67, 78, 80.
[36] Bulwer-Lytton's annotations to his copy of *Glenarvon* criticize the style as 'at once meagre and inflated'.

Not the least part of that growing scepticism involved her attitude towards revolution. In early 1848 she wrote to her old teacher
Margaret Wooler, reflecting on the change in her politics over the
last decade:

I have now outlived youth; and though I dare not say that I have outlived
all its illusions—that the romance is quite gone from Life, the veil fallen
from Truth, and that I see both in naked reality—yet certainly many things
are not to me what they were ten years ago; and among the rest, the 'pomp
and circumstance of war' have quite lost in my eyes their factitious glitter—
I have still no doubt that the shock of moral earthquake wakens a vivid
sense of life both in Nations and individuals; that the fear of dangers on a
broad national scale diverts men's minds momentarily from brooding over
small private perils, and, for a time, gives them something like largeness of
views; but, as little doubt have I that convulsive revolutions put back the
world in all that is good, check civilisation, bring the dregs of society to its
surface, in short it appears to me that insurrections and battles are the
acute diseases of nations, and that their tendency is to exhaust by their
violence the vital energies of the countries where they occur. That England
may be spared the spasms, cramps, and frenzy-fits now contorting the
Continent and threatening Ireland, I earnestly pray![37]

The terms are remarkably close to those used by George Man
Burrows and William Saunders Hallaran to express the mental and
physical dangers of revolution (see Chapter 4). Sally Shuttleworth
has shown that Brontë had an intense interest in the progress of
medical theory and practice, reading widely in the journals and
books available at the Library of the Keighley Mechanics Institute
and attending occasional lectures on physiology and related subjects. It is not unlikely that she had read Burrows and Hallaran, or
other medical commentators on revolution. Alternatively, she may
have heard them discussed by her father, who also had a keen and
critical interest in contemporary medicine.[38] The reference to the
findings of contemporary physicians is indirect in the letter, but it
implicitly supports Charlotte Brontë's move away from a romantic

[37] Charlotte Brontë, letter to Margaret Wooler, 31 Mar. 1848, in Thomas James
Wise and John Alexander Symington (eds.), *The Brontës: Their Lives, Friendships
and Correspondence* (The Shakespeare Head Brontë), 4 vols. (Oxford, 1933), ii.
202–3.
[38] See Sally Shuttleworth, ' "The Surveillance of a Sleepless Eye": The Constitution of Neurosis in *Villette*', in George Levine (ed.), *One Culture: Essays in Science
and Literature* (Madison, Wis., 1987), 316, and see her forthcoming book on
Charlotte Brontë and 19th-cent. psychology.

view of rebellion to a more sober and mature 'realism'. *Jane Eyre*, too, uses the insights of mid-nineteenth-century medical writing to support its recasting of Romantic ideas about insurrection and insanity.

When Jane Eyre, new to her post as governess at Thornfield Hall, first sees the servant Grace Poole, whom she believes is responsible for the strange laughter heard from time to time in the upper apartments, the 'romantic reader' takes a cold dose of 'plain truth' (133). The 'curious laugh' emanates from one of the rooms off the long, dark, narrow passage 'with its two rows of small black doors all shut, like a corridor in some Bluebeard's castle' (129). A 'distinct, formal, mirthless laugh', it begins 'very low' but crescendos into 'a clamorous peal that seemed to echo in every lonely chamber': 'the laugh was as tragic, as preternatural a laugh as any I ever heard; and, but that it was high noon, and that no circumstance of ghostliness accompanied the curious cachination; but that neither scene nor season favoured fear, I should have been superstitiously afraid' (130). The materials are strongly reminiscent of Bulwer-Lytton, but, in a pattern found repeatedly in *Jane Eyre*, syntax staves off fear. Danger is there, threatening, but held at a distance, made to wait its articulation until Jane has detailed the case against alarm. Even then alarm is prefaced and branded by the label 'superstition': 'but that . . . and that . . . but that . . . I should have been superstitiously afraid'. When Grace Poole appears, the uncanny is ruled out by the overwhelming evidence of physiognomy. Jane starts at the gap between the tragic sound and the woman now in front of her. Grace is 'a set, square-made figure, red-haired, and with a hard, plain face . . . any apparition less romantic or less ghostly could scarcely be conceived'. Nature clashes with the preternatural, tragedy with the everyday, ghostliness with high noon.

Similar tensions make themselves felt repeatedly in Brontë's approach to Bertha Mason. For a start, Bertha is out of time with the rest of the novel. Locked away in the attic of Thornfield Hall for ten years, along with all the outmoded furniture, she has seen no one but her stolid keeper and, occasionally, the husband she hates. In terms of her function in the plot (though not in other respects) critics are right to suggest that she is correspondingly outmoded[39]—

[39] Armstrong, *Desire and Domestic Fiction*, 208–11; and Gilbert and Gubar, *The Madwoman in the Attic*, 314, 347–8; Susan Meyer, 'Colonialism and the Figurative

a costume drama madwoman used for theatrical effect much as Mr Rochester's guests ransack the apartments on the third floor for the props and costumes of their charade party. The 'eccentric murmurs' and the 'low, slow, ha! ha!' emanating from the attic, the white figure with a mane of dark hair moving, candle in hand, through the corridors at night, and the ghostly, nightmarish appearances in the heroine's bedchamber at the darkest hours of night belong more to the apparatus of late eighteenth-century and early nineteenth-century gothic romance than to Victorian domestic realism. A remnant of the *Blackwood's Magazine* strain in the Brontës' juvenilia,[40] these traces of earlier fictional styles to some extent justify Q. D. Leavis's claim that Charlotte Brontë was 'by no means the "Victorian" product she is generally thought of as being'.[41]

Brontë's deliberate countering of romanticism with sturdy realism is nowhere more evident than in her handling of the most obvious fictional model for Bertha Mason—Sir Walter Scott's Ulrica the Saxon. Brontë's close knowledge of Scott's fiction is the one certain link between her fiction-reading on the subject of female insanity and that of Bulwer-Lytton. Writing to her friend Ellen Nussey in 1834 to recommend a course of reading, she was adamant that the canon of good recent fiction could be reduced to one author: 'For fiction—read Scott alone; all novels after his are worthless.'[42] The death of Bertha Mason is clearly modelled on chapter 31 of *Ivanhoe* (see Chapter 4 above). Like Ulrica, Bertha Mason perishes dramatically, throwing herself from the roof-top amid the flames she has lighted to destroy her tyrant and her prison. The innkeeper at the Rochester Arms describes the scene to Jane Eyre:

Strategy of *Jane Eyre*', *Victorian Studies*, 33 (1990), 247–68: 255–6. Cf. Adrienne Rich, 'Jane Eyre: The Temptations of the Motherless Woman', in *On Lies, Secrets, and Silence: Selected Prose 1966–1978* (1979; London, 1980), 97.

[40] Particularly relevant to *Jane Eyre* is the story 'Visits in Verreopolis', written in Dec. 1830 when Brontë was 14. See *An Edition of the Early Writings of Charlotte Brontë*, i: *The Glass Town Saga 1826–1832*, ed. Christine Alexander (Oxford, 1987), 316–27 and Alexander, *The Early Writings of Charlotte Brontë* (Oxford, 1983), 246. Susan Meyer has discussed Bertha Mason in relation to Brontë's juvenile writing about a black West Indian rebel, Quashia Quamina, who, in the style of the Demerara uprising of 1823, leads revolutions against her white colonists. 'Colonialism and the Figurative Strategy of *Jane Eyre*'.

[41] Q. D. Leavis, Introduction to *Jane Eyre* (London, 1966), 8.

[42] Wise and Symington (eds.), *The Brontës*, i. 122. On the Brontë's childhood reading, see Alexander, *Early Writings of Charlotte Brontë*, 11–26.

'. . . she was standing, waving her arms, above the battlements, and shouting out till they could hear her a mile off: I saw her and heard her with my own eyes. She was a big woman, and had long, black hair: we could see it streaming against the flames as she stood. I witnessed, and several more witnessed Mr. Rochester ascend through the skylight on to the roof: we hear him call "Bertha!" We saw him approach her; and then, ma'am, she yelled, and gave a spring, and the next minute she lay smashed on the pavement.'

'Dead?'

'Dead! Aye, dead as the stones on which her brains and blood were scattered.' (548)

In outline, the incident is close to Scott, but the mode of presentation is strikingly different. Ulrica dies at the height of the Saxon–Norman battle, with a song of victory on her lips. When Bertha Mason leaps to her death, her shouts can be heard a mile off, but her words—if they are words—are not recorded for posterity. The scene is described to Jane some months after its occurrence, by a 'respectable-looking middle-aged man', formerly butler to the Hall, whose eagerness to affirm his reliability as a witness gives the account a sometimes comic undertow—'I saw her and heard her with my own eyes.' In *Ivanhoe*, Ulrica is last seen 'tossing her arms abroad with wild exultation, as if she reigned empress of the conflagration which she had raised', and when the turret finally gives way, her death is simply stated—'she perished in the flames which had consumed her tyrant' (346). Bertha's end is, by comparison, brutally realized: she leaps with a yell to her death, and the impact scatters her brains and blood across the pavement. The innkeeper apparently takes some pride in the accuracy of his simile: Mrs Rochester was as 'dead as the stones' on which she lay smashed.

Scott's influence is thus clearly invoked but, at the same time, held firmly at a distance, and that caution colours all Brontë's descriptions of Bertha Mason. The description of Rochester's wife when he finally reveals her is a famous one:

In the deep shade, at the further end of the room, a figure ran backwards and forwards. What it was, whether beast or human being, one could not, at first sight, tell: it grovelled, seemingly, on all fours; it snatched and growled like some strange wild animal: but it was covered with clothing; and a quantity of dark, grizzled hair, wild as a mane, hid its head and face. (370)

This is, as many critics have argued more or less accusingly, no humane vision of female insanity. Only once does Jane Eyre make a plea for Bertha's right to the title of an 'unfortunate lady', and she is promptly assured by Rochester that sympathy is unwarranted. For the rest, Bertha is heard 'snarling' and 'snatching, almost like a dog quarrelling', laughing her eerie laugh, and (when she attacks her brother) emitting one 'fearful shriek' which 'not the wildest-winged condor on the Andes' could produce twice in succession (262, 258). 'Some strange wild animal', a 'wolfish' thing, a 'dog', a 'tigress', a 'clothed hyena', Bertha elicits a wide range of zoological comparisons (370, 392, 262, 267, 370). The emphasis, evidently, is on the madwoman's reduction to bestiality, but there are also important subtexts to the choice of terms.

Perhaps the richest involves the repeated comparison with the hyena. Pliny made the following observations on the hyena in his *Natural History*: 'The Hyena is popularly believed to be bisexual and to become male and female in alternate years, the female bearing offspring without a male but this is denied by Aristotle.'[43] The animal's perceived sexual ambivalence and its association with female revenge continued through to the nineteenth century. Milton's Samson screams at Dalila 'Out, out, hyena! These are thy wonted arts, And arts of every woman false like thee' (line 748). In 1795 Horace Walpole famously denounced Mary Wollstonecraft as 'that hyena in petticoats', 'daily discharg[ing] her ink and gall on Marie Antoinette, whose unparalleled sufferings have not yet staunched that Alecto's blazing ferocity'.[44] *Household Words* summed up the hyena's associations for mid-nineteenth-century readers a few years after the publication of *Jane Eyre* when Dickens published a comically elaborate study of the animal, in which the reasons for its poor historical reputation were painstakingly elaborated, and its liking for human flesh and taste for

---

[43] Pliny, *Natural History*, VIII. xliv (Loeb ed., iii. 77). This reference and the one to *Samson Agonistes* are indebted to Valerie Grosvenor Myer, '*Jane Eyre*: The Madwoman as Hyena', *Notes and Queries*, 233 (1988), 318. Myer does not discuss the hyena's political associations.

[44] Horace Walpole, letter to Hannah More, 24 Jan. 1795, *The Letters of Horace Walpole, Fourth Earl of Orford*, ed. Mrs Paget Toynbee, 16 vols. (Oxford, 1905), xv. 337–8. As late as 1884 the term 'political hyenaism' was being used to describe the danger of an outbreak of socialist violence. *Blackwood's Edinburgh Magazine*, 136 (1884), 210.

cannibalism grudgingly conceded. A developing friendship with the hyena at London Zoo had nevertheless convinced the writer that this maligned animal possessed a touching capacity for doglike affection.[45]

Rather than the madwoman licensing the portrayal of revolutionary desire, as she does in 1810s fiction, in *Jane Eyre* female insanity brings no 'factitious glitter'. Though her role in the plot clearly alludes to Scott's Romantic model, the vocabulary applied to her draws on traditions which are deeply hostile to Romanticism. The strength of that hostility has rarely been recognized. The most overt connection of Bertha Mason with political radicalism, for example, occurs in a passage which has absorbed a heavy share of critical attention: the passionate outburst against the restricted horizons of women's lives which ends with the madwoman's laughter (132–3). This was the moment of 'deformed' and 'twisted' rage which so upset Virginia Woolf,[46] and it is no wonder that Elizabeth Rigby felt threatened. Few passages of nineteenth-century fiction more unambiguously connect a woman's resentment at the circumscription of her abilities to the spirit of political radicalism—or more carefully differentiate the two forms of rebellion ('Nobody knows how many rebellions *besides* political rebellions ferment in the masses of life which people earth'; 132, my emphasis). The vast majority of recent critics have read the passage in terms of a feminist conviction that madness represents a mode of rebellion against the constraints of patriarchy, so that Bertha's laughter gives voice to Jane's intense resentment of the restraints placed upon her.[47] Yet to read the novel this way involves overruling Brontë's determination not to give female insanity the endorsement—or even the ambiguity—of a Romantic reading. For Lady Caroline Lamb or, indeed, for Scott, Bertha's insanity would have been an acceptable expression of the heroine's frustrated desire for all the 'incident, life, fire, feeling' denied her in her 'actual existence' (132). For Brontë, Bertha's laughter indicates the acute psychological danger of this restless discontent that agitates her heroine 'to pain'.

[45] 'Justice to the Hyæna', *Household Words*, 1 Jan. 1853, 373–7.
[46] Virginia Woolf, *A Room of One's Own* (London, 1929), 104.
[47] e.g. Rich, 'Jane Eyre', 97–8; Gilbert and Gubar, *The Madwoman in the Attic*, 349, 360; Valerie Grosvenor Myer, *Charlotte Brontë: Truculent Spirit* (London, 1987), 109–10.

I argued in Chapter 4 that the convention of the love-mad woman became enlisted most efectively in fiction about revolution not in the 1790s and early 1800s when the possibility of an over-throw of the English government was strongest but in those later years when the political situation, though troubled, seemed less unstable. Brontë's writing about female rebellion similarly belongs within a predominantly conservative context where the desire to do justice to the hungering of 'millions' after freedom from restraint contends with a deep suspicion of radicalism.[48] But, whereas Maturin, Lamb, and Scott work within the basic terms of the love-madness convention, Brontë's concern not to romanticize either rebellion or insanity leads to a far more thoroughgoing hostility to sentimental and Romantic models. The result is something like a double edge to Brontë's depiction of female madness, by which she exploits her readers' awareness of the recent tradition of associating political insurrection with female insanity, and at the same time strives to alter what that association is allowed to say. It is in this respect that *Jane Eyre* differs most fundamentally from *Lucretia*, despite the many parallels between them. *Lucretia* offers no anti-dote to its madwoman; nobody withstands the evil eye of its mon-strous 'child of night', and the past holds the present in a remorseless grip. *Jane Eyre* by contrast, respects the past, but resists its power to shape the present.

In order to do so, the novel has recourse to a series of concepts current in mid-nineteenth-century medical thinking about the mind and its pathologies. Although Bertha is a theatrically outmoded agent in the plot of *Jane Eyre*, she simultaneously fits a much more current analysis of insanity, and it is here, in the juxtaposition of the out-of-fashion with the contemporary, that Brontë's careful melding of realism with gothic supernaturalism does its most im-portant work. The melodrama of Bertha's moonlit walks is of a very different order from her actual appearance when we finally see the face of madness in this novel. Bertha's insanity is shockingly out of key even with the murderous mania of Lucy Ashton or Ulrica. The ghostly figure who leans over Jane Eyre two nights before her

---

[48] On conflicting political sympathies in *Jane Eyre*, see Terry Eagleton, *Myths of Power: A Marxist Study of the Brontës* (London, 1975), 15–32. See also Brontë's much-quoted letter on 'this ill-advised movement', in which she adds that 'their grievances should not be . . . neglected, nor the existence of their sufferings ignored'; Wise and Symington (eds.), *The Brontës*, ii. 203.

intended marriage to Mr Rochester is '[f]earful and ghastly' she tells him:

'—oh, sir, I never saw a face like it! It was a discoloured face—it was a savage face. I wish I could forget the roll of the red eyes and the fearful blackened inflation of the lineaments!'
  'Ghosts are usually pale, Jane.'
  'This, sir, was purple: the lips were swelled and dark; the brow furrowed: the black eye-brows wildly raised over the blood-shot eyes.'   (358)

Gayatri Spivak, Penny Bouhmela, and Susan Meyer have all drawn attention to the racist undertones of the Brontë's descriptions of Mrs Rochester: the rolling eyes and blackened features of the 'savage face' recalling contemporary descriptions of colonized Africa.[49] Part of the repertoire of nineteenth-century medical and anthropological writing about the 'savage races' was a conviction of their predisposition to alcoholism (Rochester tells Jane, later, that intemperance loomed large among his wife's vices, and it is one of several links between her and her keeper, Grace). It is therefore the more significant that the description of Bertha's face chimes so closely with the demented recollections of Jane's childhood persecutor, her Aunt Reed, a few pages earlier in the novel. The dying woman's mind goes back to her drunken and debauched son John, whose suicide has wrecked her health: 'He threatens me—he continually threatens me with his own death, or mine; and I dream sometimes that I see him laid out with a great wound in his throat, or with a swollen and blackened face' (291). Bertha Mason's face is similarly a blackened, swollen face of debauchery and alcoholism.[50] The fact that the apparition reminds Jane of nothing so much as 'the foul German spectre—the Vampyre' (358) accords with folkloric belief that alcoholics were 'prime candidates for revenants'. In Eastern European tradition, the vampire was not the pale creature

[49] On Bertha and the representation of race in *Jane Eyre*, see Gayatri Chakravorty Spivak, 'Three Women's Texts and a Critique of Imperialism', *Critical Inquiry*, 12 (1985), 243–61, esp. 247–9; Meyer, 'Colonialism and the Figurative Strategy of *Jane Eyre*'; Penny Boumelha, *Charlotte Brontë* (Hemel Hempstead, 1990), 58–77. Meyer notes the links between Bertha Mason and Blanche Ingram; 'Colonialism and the Figurative Strategy of *Jane Eyre*', 259–60.

[50] A number of writers have suggested that Charlotte Brontë was drawing on her experience of her brother Branwell's increasing dependency on alcohol and drugs during the time that she was writing *Jane Eyre*. For a summary, see Cynthia A. Linder, *Romantic Imagery in the Novels of Charlotte Brontë* (London, 1978), 60–1.

of horror films, but 'florid, or of a healthy color, or dark', the body swollen, the eyes open and sometimes reddened from the blood consumed. As Paul Barber notes, heavy drinkers, who acquire a ruddy complexion from the distension of their capillaries by alcohol, 'may be compared to the vampire'.[51] It is a parallel Bertha has already justified by biting into the shoulder of her brother, Richard Mason, sucking his blood, and threatening to 'drain [his] heart' (267).

The revulsion excited by Bertha Mason in all but the inured Grace Poole is entirely in line with the disgust which many medical writers of the period expressed unapologetically towards a certain class of lunatic, even at a time when Victorian medicine was priding itself on a more enlightened and paternalistic approach to the insane. Bertha Mason falls clearly within J. C. Prichard's category of 'moral insanity'.[52] The general concept of moral derangement had been floating around in medical textbooks since at least the late eighteenth century.[53] Esquirol had given it new prominence in his influential treatise on insanity, published in the late 1830s, and translated into English in the mid-1840s,[54] but it was left to Prichard (Esquirol's chief popularizer in England) to put 'moral insanity' on the English medical map as a distinctive category of mental illness. Prichard was Physician to the Bristol Royal Infirmary and to St Peter's Hospital, Bristol, at the time. His account of moral insanity was published in two forms in 1833: first in Tweedie, Forbes, and Conolly's *Cyclopaedia of Practical Medicine*, then as a monograph, *Treatise on Insanity*. Prichard described moral insanity as 'a morbid perversion of the natural feelings, affections, inclinations, temper, habits, and moral dispositions, without any notable lesion of the intellect or knowing and

---

[51] Paul Barber, *Vampires, Burial, and Death: Folklore and Reality* (New Haven, 1988), 29, 41.

[52] This point is made in Peter Grudin, 'Jane and the Other Mrs Rochester: Excess and Restraint in *Jane Eyre*', *Novel*, 10 (1977), 145–57. See also Boumelha, *Charlotte Brontë*, 61, and cf. Philip Martin, *Mad Women in Romantic Writing* (Brighton, 1987), 124–39.

[53] e.g. Thomas Arnold, *Observations on the Nature, Kinds, Causes, and Prevention of Insanity, Lunacy or Madness*, 2 vols. (Leicester, 1782–6), i. 89; and Benjamin Rush, *An Inquiry into the Influence of Physical Causes upon the Moral Faculty* (Philadelphia, 1786).

[54] Jean-Étienne Dominique Esquirol, *Des maladies mentales* (Paris, 1838); trans. as *Mental Maladies: A Treatise on Insanity*, trans. with additions by E. K. Hunt (London, 1845).

reasoning faculties and particularly without any maniacal hallucination'.[55]

The founding rationale here was to be found in one of the most evasive, yet important, concepts underlying Victorian theories of the mind: the will. 'Although associated with the highest nervous centres of the cerebral cortex, presumably in the frontal lobes, the will was generally interpreted as a mental faculty, not reducible to physiological attributes.' Indeed, 'will' was often used interchangeably with 'mind', since, in its loosest sense, it simply denoted the active deployment of mental energies.[56] Women and 'savages' were generally held to be the more vulnerable to this type of derangement, since in both the will was held to be notoriously weak.[57] As Prichard's definition of moral insanity suggests, this was one of the most pliable and potentially abusable categories in the Victorian physician's vocabulary. Prichard regaled his readers with accounts of initially mild moral disorders turning to full-blown mania before the eyes of distressed relatives:

Violent gusts of passion breaking out without cause, and leading to the danger or actual commission of serious injury to surrounding persons, are the features of the disease in most of the cases mentioned by Pinel. . . . There are other instances in which malignity has a deeper die. The individual, as if actually possessed by the demon of evil, is continually indulging enmity and plotting mischief, even murder, against some unfortunate victim of his malice.[58]

Charlotte Brontë's familiarity with the concept of moral insanity is known from a letter she wrote to W. S. Williams (her reader at Smith, Elder) a few months after the publication of *Jane Eyre*, responding to complaints from Leigh Hunt and the popular

---

[55] J. C. Prichard, 'Insanity', in J. Forbes *et al.*, *The Cyclopaedia of Practical Medicine*, 4 vols. (London, 1833–5), ii. [826].

[56] Janet Oppenheim, *'Shattered Nerves': Doctors, Patients, and Depression in Victorian England* (New York, 1991), 43.

[57] On 19th-cent. belief in the weakness of the female will, see ibid. 181–2. As a leading early 19th-cent. exponent of monogenism (the belief that the human race originated in one family) and a committed opponent of slavery, Prichard was not a likely exponent of the racist applications of the concept of moral insanity. However, the 1840s marked a crucial period in the shift towards polygenist arguments. See Nancy Stepan, *The Idea of Race in Science: Great Britain 1800–1960* (London, 1982), 1–46. Ironically, Prichard's mental psychology would be used to defeat his racial theories.

[58] Prichard, 'Insanity', [829].

novelist Julia Kavanagh that Bertha Mason was unnaturally hideous:

The character [of Bertha] is shocking, but I know that it is but too natural. There is a phase of insanity which may be called moral madness, in which all that is good or even human seems to disappear from the mind and a fiend-like nature replaces it. The sole aim and desire of the being thus possessed is to exasperate, to molest, to destroy, and preternatural ingenuity and energy are often exercised to that dreadful end. The aspect in such cases, assimilates with the disposition; all seem demonised. It is true that profound pity ought to be the only sentiment elicited by the view of such degradation, and equally true is it that I have not sufficiently dwelt on that feeling; I have erred in making *horror* too predominant. Mrs Rochester indeed lived a sinful life before she was insane, but sin is itself a species of insanity: the truly good behold and compassionate it as such.[59]

Barring the last two sentences, the letter is remarkably close in diction and content to contemporary medical writing—and particularly to Prichard, echoing his account of extreme cases in which the individual seems 'actually possessed by the demon of evil'. If a further link is wanted, it may be found in the preoccupation with fire in Prichard's case-studies of moral insanity: a young man who set out for Bishop-Thorpe Palace one day, intending to set it ablaze; a young woman paranoiacally convinced that her host is planning to fill the house with combustibles in order to destroy it.[60]

When Rochester, his attempt at bigamy thwarted, tells Jane the history of his marriage to Bertha Mason, he reproduces virtually every symptom in Prichard's repertoire. Rochester describes a woman who carries moral and mental corruption in her blood. As soon as the honeymoon was over, he tells his quiet listener, he discovered that the mother-in-law he thought dead was 'only mad; and shut up in a lunatic asylum', Bertha's younger brother was 'a complete dumb idiot', and Rochester fully expects her sallow-faced and nervous older brother to follow suit (389). This is madness carried through the female line, and expressing itself with particular virulence in women (one of the 'feminine' aspects of Bertha's madness is that her sense of time is cyclical, bound by the periodicity of

---

[59] Letter of 4 Jan. 1848, in Wise and Symington (eds.), *The Brontës*, ii. 173–4.
[60] Prichard, 'Insanity', [829], [832].

the moon[61]). Rochester finds in his 'tall, dark, and majestic' wife (389), a nature 'wholly alien to mine':

> her tastes obnoxious to me; her cast of mind common, low, narrow, and singularly incapable of being led to anything higher, expanded to anything larger . . . whatever topic I started, immediately received from her a turn at once coarse and trite, perverse and imbecile . . . her character ripened and developed with frightful rapidity; her vices sprung up fast and rank: they were so strong, only cruelty could check them; and I would not use cruelty. What a pigmy intellect she had—and what giant propensities! How fearful were the curses those propensities entailed on me! Bertha Mason,—the true daughter of an infamous mother,—dragged me through all the hideous and degrading agonies which must attend a man bound to a wife at once intemperate and unchaste.   (390–1)

Rochester has every reason to exaggerate his wife's viciousness, but it is important to remember that the weight of the novel is behind him. There is nothing in what we hear from others, or from Jane herself, to qualify either this catalogue of symptoms or the case history Rochester provides. 'Common', 'low', 'narrow', 'vicious' in her propensities and devoid of intellectual force, 'intemperate and unchaste', the 'true daughter of an infamous mother'—Bertha presents pathology of mind as an unchecked progress of decay, a cumulative disintegration of faculties unable to generate their own resistance. In her, the body becomes a mere compilation of its own vices and excesses, without the possibility of reparation or even regret, and therefore without any mechanism for withstanding the degradations of the past.

   Above all it is essential that Rochester convince Jane, and that the novel convince us, that Bertha never loved her husband. In fact, as Jean Rhys picked up adeptly in *Wide Sargasso Sea* (1966) (and as Charlotte Brontë indicated in the letter to Williams), *Jane Eyre* gives us almost no access to Bertha's feelings. We know only that Rochester never loved: that he was 'dazzled, stimulated: my senses were excited; and being ignorant, raw, and inexperienced, I thought I loved her' (389). In Bertha Mason the convention of love-madness is invoked as bitter parody. Following a black version of the sentimental stereotype, her insanity increases as her husband registers his own loathing for her and his desire to abandon the marriage.

---

[61] Elaine Showalter, *The Female Malady: Women, Madness, and English Culture 1830–1980* (London, 1987), 67.

Her 'hyen laugh'[62] mocks lovers' laughter, her embrace is savage, her kiss is life-threatening.

Jane Eyre's engagement with contemporary theories of the will goes well beyond a diagnostic model for interpreting its madwoman. Indeed, the novel's most weighty statements about the physiology and pathology of the mind occur not in relation to Bertha but in connection with Jane. To read Bertha Mason as the heroine's dark double (an interpretation which has dominated criticism of the novel since *The Madwoman in the Attic*) is to distort a book which so emphatically asserts their alterity. The taste, current in earlier romantic fiction, for symbolic alliances between a sympathetic female lead and a mad 'other woman' (see Chapter 4) is invoked only to be firmly repudiated. Rochester insists that he was drawn to Jane in the first place because 'there was a pleasureable illumination in [her] eye . . . which told of no bitter, bilious, hypochondriac brooding' (399). He demands that the reader compare Bertha Mason with Jane Eyre: '"That is *my wife*," said he. ". . . And *this* is what I wished to have. . . . Compare these clear eyes with the red balls yonder—this face with that mask—this form with that bulk"' (371). When Gilbert and Gubar read these lines they see them as forging a link between the outwardly composed governess and the raging madwoman, but the novel asks us to take Rochester at his word: to look at the two women in order to find in them an absolute difference.

The determining feature of that difference is the strength of Jane's will. In chapter 19, Rochester, disguised as a gypsy, contrives a private interview with Jane, in the course of which he delivers a phrenological reading of her head.[63] He discerns a passionate sensibility struggling with pride and reserve. Passion, he anticipates, would win were it not for the brow:

that brow professes to say,—'I can live alone, if self-respect and circumstances require me so to do. I need not sell my soul to buy bliss. I have an

---

[62] Rosalind to Orlando, in *As You Like It*, IV. i. 149–56: 'I will be more jealous of thee than a Barbary cock-pigeon over his hen . . . I will weep for nothing, like Diana in the fountain, and I will do that when you are dispos'd to be merry. I will laugh like a hyen, and that when thou art inclin'd to sleep.'

[63] There is an extensive literature on this subject. See particularly Ian Jack, 'Physiognomy, Phrenology and Characterization in the Novels of Charlotte Brontë', *Brontë Society Transactions*, 15 (1970), 377–91; and Karen Chase, *Eros and Psyche: The Representation of Personality in Charlotte Brontë, Charles Dickens, George Eliot* (London, 1984), 54–9.

inward treasure, born with me, which can keep me alive if all extraneous delights should be withheld; or offered only at a price I cannot afford to give.' The forehead declares, 'Reason sits firm and holds the reins, and she will not let the feelings burst away and hurry her to wild chasms. The passions may rage furiously, like true heathens, as they are; and the desires may imagine all sorts of vain things: but judgement shall still have the last word in every argument, and the casting vote in every decision. . . .' Well said, forehead . . . (252)

Rochester's phrenological analysis of Jane's character pivots on the words 'will' and 'shall'. This is a mind which 'declares' its own capacity for reason in its very bones. The passion interpolates and places clear bounds upon a Romantic view of insane passion, overruling all restraint.[64] The language of colonialism (heathen passions subject to the control of an imperial reason) sets up a racial model of psychology which will be explored more literally through St John later. The gypsy scene is curiously slanted, however, by the reader's gradual or, perhaps, retrospective recognition that a kind of seduction is going on here. Phrenology licenses talk of passion and love, and when Rochester speaks of Jane's controlling will he is, at least initially, inviting its relaxation. Echoing Keats's famously narcotic 'Ode to a Nightingale', Jane registers her own bedazzlement as Rochester abandons his disguise: 'Where was I? Did I wake or sleep? Had I been dreaming? Did I dream still?' (253). But Jane's will is capable of withstanding enfeeblement. She congratulates herself, at the end, on having 'fallen into no great absurdity' (254), and Rochester is compelled to acknowledge that his phrenological subject has defeated him.

Love itself emerges, in *Jane Eyre*, as a battle of wills. The time-honoured diction of lovers, in which to desire another is to be without control over oneself, is acutely problematic for Jane because it involves her in a state of mental incapacity which seems to her to lie on a continuum with insanity. That derangement may be brought on by thwarted desire is, of course, part of the familiar rhetoric of lovers, and it lies at the heart of the convention of the love-mad woman. But Brontë's unflinching application of contemporary theories of the mind to that tradition means that Jane's love for Rochester becomes a state akin to madness, even

---

[64] On Brontë's combination of an Enlightenment tradition of philosophical psychology with a more recent Romantic temptation to resolve mental conflicts into agonized dualisms, see Chase, *Eros and Psyche*, 52–65.

before the necessity of losing him. Jane fears Rochester's love as the destruction of her will. 'He made me love him' (219), she reflects just before the gypsy scene, invoking the language of phrenology and mesmerism in her turn: 'My master's colourless, olive face, square, massive brow, broad and jetty eyebrows, deep eyes, strong features, firm, grim mouth,—all energy, decision, will . . . were full of an interest, an influence that quite mastered me,—that took my feelings from my own power and fettered them in his' (218). And later, seeing him waiting for her when she returns from her visit to the dying aunt: 'Well, he is not a ghost; yet every nerve I have is unstrung: for a moment I am beyond my own mastery' (306). Even here, the resistance is palpable: only 'for a moment' is Jane's mind out of her own control. She fights at every turn this psychological equivalent to her earlier social complaint voiced against the background of the madwoman's laughter, this reduction of the ability to exercise her own faculties and pursue her own efforts. The key factor in Jane's reaction to the disclosure that Rochester has a wife is dismay not at the collapse of her marriage hopes but at the inequality his deception forced upon her. In not having comprehended her own circumstances and exerted her will on a correct basis, Jane sees herself as self-alienated, dangerously reduced, and morally sickened.

In this context Rochester's claim that his love for Jane would not fail if she were to go mad—that his arms would confine her, not a strait waistcoat; and that his embrace would be 'at least as fond as it would be restrictive' (384)—is self-defeating. The fact that he can speak the language of the mad-doctor as the language of love is the final proof for Jane that such love would indeed subdue her to a state of moral dependency akin to insanity. The comparison seems excessive, but the moral and psychological terms in which Charlotte Brontë deals require that it should be so. When Rochester tries to persuade her to renounce conventional morality and live with him as his wife, Jane conceives of her choice as one between sanity and madness: spurning the cries of Feeling, she insists 'I will hold to the principles received by me when I was sane, and not mad—as I am now' (404). Lurking behind Jane's response to the disclosure of her lover's marriage is the presence of the love-madness convention, but it has become almost unrecognizable. The temptation to 'slip into the cliché so conveniently provided by fiction' (to borrow

Virginia Woolf's phrase[65]) is simultaneously registered and rejected as Jane's will insistently makes itself heard. The talismanic phrase 'I will' dominates her narrative at this point more than ever, demanding the return of the traitor Reason. Indeed, 'I will' becomes something close to a performative utterance, in so far as energetic expression implies the active engagement of the mind.

The achievement of sanity is not, however, without some cost or some rhetorical difficulty. When she faces the necessity of leaving Rochester, Jane comes very close to accepting the pull of cliché. Overwhelmed with misery at her own blindness, she descends into her deepest mental crisis:

I seemed to have laid me down in the dried-up bed of a great river; I heard a flood loosened in remote mountains, and felt the torrent come: to rise I had no will, to flee I had no strength. I lay faint; longing to be dead. One idea only still throbbed life-like within me—a remembrance of God: it begot an unuttered prayer: these words went wandering up and down in my rayless mind, as something that should be whispered; but no energy was found to express them:—

'Be not far from me, for trouble is near: there is none to help.' (374)

In a reversal of Jane's characteristic patterns of speech, the 'I' is displaced, syntactically delayed in a verbal replica of her mental dilemma: 'to rise I had no will, to flee I had no strength'. The Ophelian drama is internalized at this point, but in the process it is rendered strange and threatening. This is no stream of unconsciousness, but a terrifying torrent, the allusion to Psalm 69 insisting that Jane's crisis is spiritual as much as, or even more than, it is emotional. The sense of a tortuous effort towards self-expression is reinforced by the similarity of the passage to a water-colour she had shown Rochester from her portfolio in one of their first meetings: a sinister vision of a 'drowned corpse glanc[ing] through the green water' (153). There, too, the image at once repeats and renders alien the Ophelian iconography so beloved by late eighteenth- and early nineteenth-century sentimental painters. In a thoroughly *un*-Ophelian touch, Jane's painting introduces one of Bewick's birds in the role of scavenger, holding in its beak a bracelet torn from the corpse's arm. Sentimentalism, in Charlotte Brontë's hands, suddenly exposes itself as a horror scenario.

A kind of psychological drowning does, nevertheless, occur.

[65] Virginia Woolf, *Between the Acts* (London, 1941), 20.

Most readers would expect that the prayer which wanders, inarticulate but intensely felt, in Jane's 'rayless mind' will be heard. But prayer without 'energy' behind it is inefficacious in Charlotte Brontë's theology:

It was near; and as I had lifted no petition to heaven to avert it—as I had neither joined my hands, nor bent my knees, nor moved my lips—it came: in full, heavy swing the torrent poured over me. The whole consciousness of my life lorn, my love lost, my hope quenched, my faith death-struck, swayed full and mighty above me in one sullen mass. That bitter hour cannot be described: in truth, 'the waters came into my soul; I sank in deep mire: I felt no standing; I came into deep waters; the floods overflowed me.' (374–5)

Once again, syntax attempts to stave off fear, but this time it fails. The passage between parenthetical dashes gestures towards a prayer which would have forestalled the despair lying at the end of the sentence, but 'it came'. 'Love lost' is only one part of this overwhelming bitterness, which finds its only possible articulation in surrendering its voice altogether to the biblical passage, ending appropriately with her relegation from the subject of action to its passive object: 'I came into deep waters; the floods overflowed me.'

Those words end chapter 26, and in the first edition of the novel they ended the second volume. The hiatus before Jane raises her head at the start of chapter 27 (volume iii) marks the silence that is the closest she comes to madness. But she wakes, with her will revived and adamant: 'a voice within me averred that I could do it; and foretold that I should do it' (379). *Jane Eyre* affirms its heroine's sanity, but not, finally, by the effort of her own faculties alone. The novel gives her back her reason but does so by denying her agency: that is, by redefining her will in terms of supernatural, divine influence. In so doing it provides a complex answer to the whole question of agency that has so perplexed the novel from the start. In bed that night, after hearing Mr Rochester's confession and freely forgiving him, Jane dreams that she is once again in the red room at Gateshead; but this time the light of the moon brings sanity instead of hysteria. Anticipating the visionary passages in *Shirley* and *Villette*, the moon breaks forth in a white, female form, bending its 'glorious brow', bidding her 'flee temptation', and the key word comes back in answer: 'Mother, I will' (407).

Supernaturalism licenses and empowers the operation of the will in *Jane Eyre*, and it is here that Brontë makes her most distinctive engagement with contemporary ideas about mesmerism, or animal magnetism—the highly influential theory and practice based on the belief that there existed 'invisible emanations from one person to another, similar or identical to the imponderable fluids of electricity or magnetism'. Charismatic individuals were held to be capable of healing or, more dangerously, disturbing the bodies and minds of others by exploiting these channels of influence.[66] Karen Chase has argued that it is only through intricate attention to the spatial model of the psyche offered by phrenology that Brontë finds a psychological language subtle enough for her purposes.[67] Yet the novel's engagement with mesmeric theory is at least as important. Brontë was clearly not afraid to plunder medical and scientific theory for what she wanted, leaving behind what she did not like— particularly when it came to theology. Among the most vigorous and proselytizing spokesmen for mesmerism in the 1830s and 1840s were men like Thomas Arnold, Richard Whately, and Henry and Samuel Wilberforce, who yoked an understanding of mesmeric phenomena with theology to produce an influential account of the workings of the divine spirit. Here, God's will was likened to the power of the mesmerist to infuse and control the mesmeric subject. In the service of an evangelical mission, this was potent material. The 'mesmeric cure of souls', to use Alison Winter's phrase, was particularly rife in the North of England among the 'abundant shower of curates' whom Charlotte Brontë parodies in the opening words of *Shirley*.[68] Theological proponents of mesmerism often came into conflict, however, over their interpretation of the will. As Winter explains, mesmerists like the Anglican evangelist Thomas Pyne 'construed the subjection of the patient's will to the mesmer- ist's as an aid in missionary work', but 'it was common for mes- meric opponents to warn of the dangers of trusting oneself to another, or of weakening one's will in any respect. Given evangeli- cal emphasis on the personal struggle to become, and remain, saved, it should come as no surprise that many argued that with the weakening of the will went the decreased ability to determine one's

[66] For the origins of mesmeric theory, and its place in 19th-cent. thinking, see Alison Winter, ' "The Island of Mesmeria": The Politics of Mesmerism in Early Victorian Britain', doctoral dissertation, University of Cambridge, 1992; the defi- nition given here is taken from p. 3. See also Robert Darnton, *Mesmerism and the End of the Enlightenment in France* (Princeton, 1968).

[67] Chase, *Eros and Psyche*.

[68] Winter, 'The Island of Mesmeria', 47.

spiritual fate.'[69] In that context, it becomes clear that Brontë's enlistment of mesmerism involves a highly selective appropriation of only those elements which were compatible with her stringent commitment to a personal spiritual struggle. As her later fight with St John Rivers shows very clearly, Brontë's heroine accepts only the direct mesmeric influence of God's will, rejecting any mediation. The vision in her bedroom at Thornfield Hall accordingly acts as a strengthening of the mind through mesmeric inspiration rather than its pacification through mesmeric mastery.

Jane's narrative is from this point on primarily concerned with completing the restoration of her will by placing her in situations where she can confront and come to terms with her own history— a process already begun when she forgave her dying Aunt Reed. Among the most important aspects of the extended time Jane spends at Moor House is the renewed emphasis on the value of teaching. Almost everyone Jane encounters at Moor House is a teacher. Diana and Mary Rivers both earn their living as governesses until Jane rescues them from their drudgery; but they also tutor Jane in German and help her in a course of reading, and she, in turn, teaches them drawing. St John studies continually, and, in the hope of persuading Jane to go to India with him as his missionary wife, instructs her in the rudiments of Hindustani. Jane herself earns her living by teaching at the little church school in the village, and even the wealthy daughter of the local factory owner funds the school and visits it regularly.

This emphasis on the absorption and communication of learning is more than just a part of familial and friendly life. It takes on considerable moral importance in a novel centrally concerned with the health or sickness of the mind. Good teaching, in *Jane Eyre*, is the opposite of the tyranny she experienced as a child at Lowood School. It instils and nurtures the intellectual and moral precepts on which a healthy mental state must be based. When Jane teaches at the village school, her little pupils initially seem to her Berthas in the making: 'Wholly untaught, with faculties quite torpid, they seemed to me hopelessly dull; and, at first sight, all dull alike: but I soon found I was mistaken' (467). This first encounter with her 'little scholars' blends almost imperceptibly into a statement about her own mental health. Hoping that the best-born among them will

[69] Ibid. 54.

possess some 'germs of native excellence, refinement, intelligence, kind feeling', she goes on: 'My duty will be to develope these germs . . . Much enjoyment I do not expect in the life opening before me: yet it will doubtless, if I regulate my mind, and exert my powers as I ought, yield me enough to live on from day to day' (458). Initially, Jane finds herself unable to place much hope in those central nineteenth-century precepts, self-control and duty, yet the next chapter finds her admitting that she was mistaken in her expectations both of her pupils and of herself. In teaching her pupils to keep their tasks regular, their persons neat, and their manners quiet and orderly, she does, against all conventional expectations, find a real measure of happiness. Only her dreams evade her control, and they become almost the sole consolation the novel allows its romantic readers that Jane is still, in private, suffering for love.

The time Jane spends at Moor House and in her village school is, then, a demonstration of the therapeutic power of the will. Most of all, it is about learning to respond correctly to a past which inevitably makes its demands upon you, but which can be brought under control by an effort of mind. As Jane affirms when she finally obliges St John to leave her, 'Where there is energy to command well enough, obedience never fails' (536–7). It is this final section of the novel which distinguishes Jane most clearly from the man she loves. While Jane strengthens her mind through rigorous self-discipline and self-education, Rochester shuts out the world, to brood on the happiness he has lost and to lament the remorseless grip of the past upon him. The symmetries in this book hold far more closely between Bertha and Rochester than they do between Bertha and Jane. Not least, Rochester's account of the moral degeneration of his Creole wife is remarkably close to his confession of his own activities in the years after she is officially declared mad. Bertha degenerated into a creature of appetite, 'intemperate and unchaste'; not so differently, Rochester admits to Jane that he spent 'ten long years' moving around Europe, vainly seeking his ideal of a woman amongst 'English ladies, French countesses, Italian signoras, and German Gräfinnen' (396). He insists that he was no Don Giovanni, accumulating women out of a yearning for moral sympathy rather than for sexual conquest, so that, once again, Bertha is invoked in a spirit of repudiation rather than sympathy: 'I tried dissipation— never debauchery: that I hated, and hate. That was my Indian

Messalina's attribute: rooted disgust at it and her restrained me much, even in pleasure. Any enjoyment that bordered on riot[70] seemed to approach me to her and her vices, and I eschewed it' (397). Nevertheless, insanity clings to Rochester in the shape of his wife, and, more subtly, in continual allusions to a potential for insanity in himself. The vocabulary he uses of Bertha attaches to himself (as it once did to Jane), even though he tries to find a separate diction for their cases: he too was 'gross', and 'grovelling', when he married her for no better reason than sensual appetite; he too can 'snarl', raise his eyebrows wildly, grind his teeth, seize Jane's arm in a painful grasp, and seem to devour her with 'his flaming glance'. In a sudden access of Bertha's animalism, he even fears that Jane will now recoil from his touch 'as if I were some toad or ape' (389, 400, 403, 405, 386).

Like his wife, Rochester is a slave to his past because he did not use his will correctly. The only 'sane' course for him in this novel would have been to acknowledge his marriage and accept its claim on him. Indeed, if anyone is in the position of the maddened lover, in this novel, it is Rochester. The innkeeper at the Rochester Arms, not knowing that the subject of his story is in front of him, tells Jane that, after she left, Rochester

grew savage—quite savage on his disappointment: he never was a mild man, but he got dangerous after he lost her. . . . He broke off acquaintance with all the gentry, and shut himself up, like a hermit, at the Hall. . . . He would not cross the door-stones of the house; except at night, when he walked just like a ghost about the grounds and in the orchard as if he had lost his senses—which it is my opinion he had . . . (547)

Reading that description it is hard not to imagine that Charlotte Brontë had in mind the example of her brother, Branwell. Branwell's physical and mental collapse after the exposure of his love affair with his employer's wife, Lydia Robinson, was the single most painful example of derangement for love that Charlotte Brontë had before her during the writing of *Jane Eyre*. Significantly, she was alone in the Brontë family in showing no sign of sympathy towards her brother. His 'refusal' to 'fight his sorrow' gave him no credit at all with Charlotte, who had so recently been forced to

[70] Like Joseph Mason Cox (p. 45 above), Rochester is using the word in the now almost obsolete sexual sense.

subdue her own impossible love for her Brussels teacher, Constantin Héger.[71]

Jane's and Rochester's stories come together again only at that point where Rochester has paid heavily for his transgressions and acknowledged God's authority. The agency of his restoration to health is, again, supernatural, but though Rochester and Jane share the uncanny experience which carries their voices to each other over hundreds of miles, they register its significance very differently. In conversation with St John, fighting his determination that she should go with him to the missionary fields of India as his wife, Jane struggles to know God's will. Whether what follows is divine inspiration or 'the effect of excitement', she leaves the reader to judge, but the pressure towards a spiritual interpretation is fairly clear:

The one candle was dying out: the room was full of moonlight. My heart beat fast and thick: I heard its throb. Suddenly it stood still to an inexpressible feeling that thrilled it through, and passed at once to my head and extremities. The feeling was not like an electric shock; but it was quite as sharp, as strange, as startling: it acted on my senses as if their utmost activity hitherto had been but torpor; from which they were now summoned, and forced to wake. They rose expectant: eye and ear waited while the flesh quivered on my bones. . . . I saw nothing, but I heard a voice somewhere cry—'Jane! Jane! Jane!'   (535–6)

Jane's experience is one of pure inspiration. For the third time in the novel her heart stands still, but her earlier experiences of near-insanity are conclusively rewritten in terms of sane inspiration. Rochester, on the other hand, is bereft of sanity at this moment. He cries out to God under the sheer pressure of an intolerable longing: 'I asked of God, at once in anguish and humility, if I had not been long enough desolate, afflicted, tormented . . . the alpha and omega of my heart's wishes broke involuntarily from my lips, in the words—"Jane! Jane! Jane!" . . . If any listener had heard me, he would have thought me mad: I pronounced [the words] with such frantic energy' (572). The difference between Rochester's 'frantic energy' and Jane's 'thrilling' alertness is a crucial distinction in the health of the mind. When Rochester recounts his version of this

---

[71] See Winifred Gérin, *Branwell Brontë* (London, 1961), esp. 245, 279. Branwell's one consolation amid his consuming sorrow was the fond (and mistaken) belief that his beloved was 'distracted to the verge of insanity' by their separation; 266.

experience to Jane, her response is revealing. She declines to return the confidence, keeping her own story to herself, telling the reader that 'the coincidence struck me as too awful and inexplicable to be communicated or discussed' (573). More to the point, she knows that Rochester, blind, maimed, and dependent (a condition she once feared would be hers, morally, as his lover), could hardly bear it: 'that mind, yet from its sufferings too prone to gloom, needed not the deeper shade of the supernatural. I kept these things, then, and pondered them in my heart' (573). The alignment of herself with the Virgin Mary in those last words reinforces, if it is needed, the superiority of her will, granting her a privileged glimpse into the workings of the divine will itself.

Only Jane has a mind equipped to bear the knowledge of the supernatural in *Jane Eyre*, but, as Gayatri Spivak has rightly insisted, the heroine's individual story is not the only one Charlotte Brontë is interested in telling. Though she cannot find a language for combining the irreconcilable demands of Jane's private happiness and the wider social mission contemplated at Moor House, the fact that the novel's conclusion is left to St John Rivers is the strongest indication of a 'soul-making' mission[72] that reaches well beyond the novel's love-story even as it depends upon the same psychological principles. But whose soul is being made? Ostensibly, the heathen's: when St John takes up his mission in India, where he 'labours for his race' and 'clears their painful way to improvement', he is literalizing that train of imagery running through the novel, in which the 'heathen passions' are subjugated to an imperial will. Love and empire appear to share the same language, as St John's death finally lifts Jane's romance on to an altogether higher plane. His last letter, closing the novel, affirms a zealot's absorption in God's will so complete that heart and mind are beyond violation: 'No fear of death will darken St. John's last hour: his mind will be unclouded; his heart will be undaunted' (578–9). Yet this is a novel which has constantly reminded its reader that the will is a hard master as well as a just one, and even as St. John's devotion of his life to God is endorsed, it is implicitly criticized. God's will, we have already been told, has operated just as surely in bringing Jane and Rochester together, and as they grow completely one with each other, the need to bring feeling under control is no longer an issue.

[72] See Gayatri Spivak, 'Three Women's Texts', 248.

The exercise of the will no longer has any place in their conclusion because judgement and feeling are at one.

Because the concept of the will is nebulous and clinically dubious in twentieth-century eyes, and yet, at the same time thoroughly familiar, it has been easy for recent criticism to discount this aspect of Brontë's writing, in favour of a post-Freudian emphasis on the disruptive work of the unconscious. Yet *Jane Eyre* is wedded to an understanding of the mind essentially hostile to the uncontrolled messages of desire, whether they be personal or social. Everything in *Jane Eyre* works towards this achievement: this ability to recall and repair the past on one's own terms, under the command of one's own will. The past, for Walter Scott, was a source of potentially radical identifications, enabling the construction of the present as a state of travesty. For Charlotte Brontë, one of Scott's most admiring readers but also, perhaps, the most determined to argue with him, identity consists in mastering that past, acknowledging it as part of one's history, but a history whose meaning is under control. Though it will always leave its scars, the wounds— like Rochester's—do, eventually, heal. Brontë's rewriting of Romantic insanity has all the confidence and yet all the anxiety of the mid-Victorian period: that acute awareness of a danger past but still lurking, mixed with considerable confidence that the threat can be forestalled. In that context, it is not surprising that the madwoman as a distorted and damaged descendant of the love-madness convention provided Brontë's novel with its most powerful symbol. She carries the memory of earlier nineteenth-century writing about female insanity and political rebellion into a context where history, like genre, can be seen to be manipulable.

6
The Woman in White, Great Expectations, *and the Limits of Medicine*

WILLIAM MAKEPEACE THACKERAY deeply admired *Jane Eyre*. Ignoring a publisher's deadline, he spent a whole day closeted with the book, and was found weeping at one of the love scenes by the manservant who came with coals for the fire.[1] Nevertheless, his feelings were more mixed when Charlotte Brontë dedicated the second edition of the novel to him, describing him effusively as one who 'comes before the great ones of society much as the son of Imlah came before the throned kings of Judah and Israel, and who speaks truth as deep, with a power as prophet-like and vital, a mien as dauntless and as daring' (36). Thackeray was to be haunted by the tribute for years. It fuelled a rumour already current in London literary circles that the mysterious 'Currer Bell' was his children's governess, seduced by Thackeray and taking pre-emptive revenge for his portrayal of her as Becky Sharp in *Vanity Fair*. Elizabeth Rigby's anonymous review picked up the story and gave it a further boost, satirically linking *Vanity Fair*, *Jane Eyre*, and the *Report for 1847 of the Governesses' Benevolent Institution*. The gossip was still circulating more than ten years later. When an inquisitive American lady accosted him at a dinner-party in 1860, and demanded to know 'is it true, the dreadful story about you and Currer Bell?', he sighed, 'Alas, Madam, it is all too true. And the fruits of that unhallowed intimacy were six children. I slew them all with my own hand.'[2]

The opportunity for wit aside, the dedication was acutely embarrassing for Thackeray. Unbeknown to Charlotte Brontë, his wife

---

[1] *The Letters and Private Papers of William Makepeace Thackeray*, 4 vols., ed. Gordon N. Ray (London, 1945–6), ii. 318–19.

[2] Gordon N. Ray, *Thackeray: The Uses of Adversity (1811–1846)* (London, 1955), 11.

had been declared insane seven years earlier, and parts of *Jane Eyre* might well have read like a *roman-à-clef*. Isabella Thackeray's madness was believed to have been a delayed reaction to childbirth. The first symptom, noted in 1840, was depression, punctuated by strange laughter. The family doctor prescribed a seaside trip, which brought only temporary relief. A few weeks later, Isabella tried to drown their 3-year-old daughter, Annie, but recovered her senses just in time to rescue her, attempting to drown herself a few days afterwards. Thackeray was persuaded that a visit to her family in Ireland might assist a recovery, but during the voyage her depression deteriorated. At one point she threw herself into the sea and floated calmly on her back for twenty minutes, 'paddling with her hands' until the ship's boat found her.[3] By the time they arrived at Cork she was completely insane. In a letter to his mother, Thackeray described a miserable night during which he could only prevent her committing suicide by tying himself to her: 'I had a riband round her waist, & to my waist, and this always woke me if she moved.'[4] After a hideous stay with his mother-in-law, who made it clear that she blamed him for her daughter's condition, he decided, with the doctor's approval, to take Isabella to Paris. There, at the end of 1840, she was committed to Esquirol's renowned Maison de Santé at Ivry, near Paris. Five years later, having exhausted the best treatment on offer in France and Germany, Thackeray brought her back to England. Horrified by the conditions in the asylum recommended by one of the Lunacy Commissioners, his friend the poet and playwright Bryan Procter, he placed her instead in the private care of a mother and daughter at Camberwell. Isabella's mania had gradually subsided into a listless melancholia which was more or less the state in which she remained for the rest of her life; she survived Thackeray by more than thirty years. After 1847 he visited her less and less, telling his mother that there was no longer any point: Isabella had become unchangingly apathetic, missing neither him nor her daughters, and not caring '2^d for anything but her dinner and her glass of porter'.[5]

Thackeray must have found *Jane Eyre* not just incidentally

---

[3] Thackeray, *Letters and Private Papers*, i. 483.
[4] Ray, *Thackeray*, 11.
[5] Thackeray, letter to Mrs. Brookfield, 14 Oct. 1858, in *Letters and Private Papers*, ii. 440.

reminiscent of his own experience but uncannily near to it. Isabella's madness, like Bertha's, was associated with irrationality in the maternal line; both women were declared mad four years into their marriages; both were at times violent and even homicidal, though for the most part passive; and both were given to manic bursts of laughter. The strangest link of all, however, arises not in connection with Bertha but in Rochester's declaration of love to Jane. Having driven her to tears with the prospect that, once he is married, she must go as a governess to 'Bitternut Lodge, Connaught, Ireland', he relents. In the passage which follows, Rochester employs a metaphor which must have painfully recalled Thackeray's literal binding of himself to his wife in Ireland:

I sometimes have a queer feeling with regard to you—especially when you are near me, as now: it is as if I had a string somewhere under my left ribs, tightly and inextricably knotted to a similar string situated in the corresponding quarter of your little frame. And if that boisterous channel [the Irish Sea], and two hundred miles or so of land come broad between us, I am afraid that cord of communion will be snapt; and then I've a nervous notion I should take to bleeding inwardly. (316–17)

Though the connections between Brontë's story and Thackeray's life were close, they were also accidental, and Thackeray could do nothing decently to dissociate himself from the novel. Faced with a *fait accompli* in the dedication, he did his best to smooth the situation, assuring Brontë's editor W. S. Williams that, however it might be misconstrued, this was 'the greatest compliment I have ever received in my life'.[6] Advised of the problem, Charlotte Brontë was torn between amazement and mortification: 'Well may it be said that fact is often stranger than fiction!', she wrote in her turn to Williams. 'The coincidence struck me as equally unfortunate and extraordinary.'[7] The letter is fraught with its author's mortified sense that her regret was 'just worth nothing at all' when it could not repair the wrong: the error of 'making horror too predominant' in the depiction of Bertha Mason[8] had escalated in seriousness and the novel now lay wide open to misinterpretation.

Thackeray's marital history made him very much the unintended

[6] *Letters and Private Papers*, ii. 341.

[7] Brontë, letter to W. S. Williams, 28 Jan. 1848, in Thomas James Wise and John Alexander Symington (eds.), *The Brontës: Their Lives, Friendships and Correspondence*, 4 vols. (Oxford, 1933), ii. 183–4.

[8] See Ch. 5.

reader of *Jane Eyre*, one likely to feel for Rochester's mad wife with
a degree of pain not expected and not easily accommodated by a
text which stresses the monstrosity of the madwoman at the expense of her claim to pity. A reader who has cause to weep not just
at the love scenes but also at the mad scenes extends the qualities of
readerly 'gentleness' beyond the point where the novel expects them
to stop. Thackeray's precise response to Bertha Mason is not recorded, but in Brontë's eyes his unhappy situation constituted a
rebuke to the novel and to herself: 'profound pity', she had told
Williams, 'ought to be the only sentiment elicited by the view of
such degradation'.[9] Under these conditions, she did not choose to
repeat her earlier appeal to established facts as a defence. The
answer must have been apparent to her: if she knew that cases like
Bertha Mason's existed, she should have been more alert to the
need for sympathy.

Brontë's anxiety about the depiction of female insanity in *Jane
Eyre* does not depend on a simple opposition between the increasingly brutal realism of medical psychology (see above, Chapter 5)
and the more sympathetic response to female insanity traditionally
encouraged by fiction. To differentiate between medicine and the
novel in such terms would be unjust to those doctors who exercised
their responsibilities in caring for the mad as humanely as possible,
and would conveniently overlook the numerous cases in which
Victorian fiction-writers were only too happy to treat the subject of
insanity as an opportunity for horror-mongering.[10] Nevertheless,
Brontë's concern that her description of Bertha Mason had ex-
cluded pity does indicate that by the late 1840s there was a mean-
ingful distinction to be made between the kinds of knowledge and
the qualities of understanding encouraged by medical and fictional
writing respectively.[11] I argued in Chapter 2 that the degree of
sympathetic identification solicited by the sentimental convention
of the love-mad woman proved increasingly difficult for medical
writers to accommodate in their search for a more theoretically
rigorous and effectively curative approach to insanity. From the
1850s it was this same awkward question of the degree of sym-

[9] Letter of 4 Jan. 1848, in Wise and Symington (eds.), *The Brontës*, ii. 173–4;
quoted more fully in Ch. 5.
[10] For a particularly grotesque example, see Thomas Peckett Prest, *The Maniac
Father; or, The Victim of Seduction. A Romance* (London, 1842).
[11] See Ch. 5.

pathy to be shown to the mad which provided a number of novelists with the basis for articulating outright opposition to the alienists' authority. Brontë's later work is a case in point. In *Shirley* (1849), and again in *Villette* (1853), she subjects contemporary psychology to far closer critical scrutiny than in *Jane Eyre*, and, though she resists the convention of the love-mad woman, she does use the example of unhappiness in love to suggest that a crisis has emerged in the capacity of medicine and fiction to represent psychological suffering. Her scepticism is most evident in *Villette*, where she accepts what medicine has to offer, but refuses to grant doctors the exclusive right to pronounce upon a case of emotional and mental suffering which is not reducible to frustrated desire and which lies outside the narrow range of their diagnostic vision. Brontë's novel complicates its resistance to medicine, however, by offering no securer insights in place of the doctor's reductive understanding of hysteria. Lucy Snowe is a difficult and unforthcoming narrator, who thwarts the reader's sympathy as much as she thwarts the understanding of the professional and amateur psychologists around her.[12]

Brontë was far from alone in her informed and combative response to mid-nineteenth-century debates about the power of the medical profession. Other novelists were similarly determined to explore the limits of medical understanding and the destructiveness of medical knowledge, and, like Brontë, many found the theme of disappointed love a useful vehicle. Geraldine Jewsbury's unjustly neglected novel *Constance Herbert* (1855) brought romance into strained relations with contemporary anxieties about hereditary insanity. The heroine makes the grim discovery that the mother she had thought dead is in fact alive, insane, and permanently committed to medical care. Constance stoically renounces her own hopes of marriage rather than risk passing on her mother's madness to future generations.[13] In more lurid vein, Shirley Brooks's *The*

[12] See Sally Shuttleworth, ' "The Surveillance of a Sleepless Eye": The Constitution of Neurosis in *Villette*', in George Levine (ed.), *One Culture: Essays in Science and Literature* (Madison, Wis., 1987), 313–35.

[13] Hereditary insanity was a subject of mounting concern in this period, as it would be for long afterward. In 1852 Charles Dickens rejected a short story on the theme, submitted to him by Wilkie Collins for publication in *Household Words*, on the grounds that it might easily offend or distress 'those numerous families in which there is such a taint'. See Kenneth Robinson, *Wilkie Collins: A Biography* (London, 1951), 73.

Gordian Knot (1860) tells the story of a medical student driven morally insane by the marriage of his beautiful cousin to another man. Once established as the family physician, he abuses his professional role in an ultimately unsuccessful attempt to poison the marriage. The novel concludes with him in an asylum, alternately 'a raving maniac and a still more pitiable, sobbing, crouching, whining idiot'.[14] Dissimilar in tone and style though they are, both *Constance Herbert* and *The Gordian Knot* depict medical psychology as deeply destructive of human relationships, and their use of the romance mode as a means of expressing resistance to medical 'expertise' was to become familiar. For most novelists writing about insanity after 1858, however, one issue overshadowed all others.

The year 1858–9 saw the first of two major 'lunacy panics' in Britain, following the exposure of numerous cases in which sane men and women had been wrongly diagnosed as insane and denied recourse to legal or other means of contesting their certification. Public outrage, vigorously represented by the Alleged Lunatics' Friend Society, was strong enough to force the establishment of a Select Committee of Inquiry.[15] Those charged with specific abuses of the system included some of the most famous names in early Victorian medicine,[16] and the scandal brought into the open many of the covert antagonisms which had made themselves felt in fictional writing about the medical treatment of insanity over the previous decade.

The lunacy panic found direct expression in the new genre of sensation fiction. Mary Elizabeth Braddon's hugely popular *Lady*

[14] Shirley Brooks, *The Gordian Knot: A Story of Good and Evil* (London, 1860), 370.

[15] See Roger Smith, *Trial by Medicine: Insanity and Responsibility in Victorian Trials* (Edinburgh, 1981), 69.

[16] See Ch. 2. The scandal was not a new one. See Peter McCandless, 'Liberty and Lunacy: The Victorians and Wrongful Confinement', in Andrew Scull (ed.), *Madhouses, Mad-doctors, and Madmen: The Social History of Psychiatry in the Victorian Era* (London, 1981), 345. An important earlier Victorian case not mentioned in McCandless is that of George Mann Burrows, arraigned in the public press and brought before the Metropolitan Commissioners in Lunacy in 1830 on suspicion of having signed a warrant for a man's incarceration without personally inspecting him. See Burrows, *A Letter to Sir Henry Halford, Bart, K.C.H., &c., Touching Some Points of the Evidence, and Observations of Counsel, on a Commission of Lunacy on Mr. Edward Davies* (London, 1830), and the *Memoir of George Mann Burrows, M.D., F.L.S., Fellow of the Royal College of Physicians, London (Extracted from the London Medical Gazette of December 11, 1846)* (London, [c.1846]).

Audley's Secret (1861), William Gilbert's *Shirley Hall Asylum; or, The Memoirs of a Monomaniac* (1863), Charles Reade's *Hard Cash* (1864), *Foul Play* (1868) (a collaborative effort with Dion Boucicault), and *A Terrible Temptation* (1871), and J. Sheridan LeFanu's *The Rose and the Key* (1871) are a few of the novels which addressed the crisis, giving imaginative force to public fears about the competence and trustworthiness of doctors who cared for the mad. Such fiction raises questions which had not surfaced before in such stark form. How justified were the claims of the medical men? Should the novel honour those claims? Or, in an age which believed passionately in the moral obligations of art and literature, should novelists preserve their own authority as public moralists by interceding in the debate over the definition and appropriate management of insanity?

Like Geraldine Jewsbury and Shirley Brooks, many sensation-writers used the romance as a vehicle for articulating their complaints against modern medicine. For polemicists like Charles Reade and Dion Boucicault, romance made for suitably poignant contrasts between a feeling hero or heroine and a heartless medical profession. For the two writers credited with inventing the sensation form, Wilkie Collins and Charles Dickens, the question of how to deal with the wrongful incarceration debate in fiction was less easily resolved. Neither Collins nor Dickens was a neutral participant in the debate over Britain's asylum laws. The lunacy panic touched them and their immediate circle of friends uncomfortably closely in 1858, and, while they had every reason to share Reade's scepticism about the alienists' profession, there were good reasons for them to desist from seizing the moral high ground on behalf of literature.

Collins and Dickens had been close friends since the early 1850s, and both had long been interested in the progress of medicine for the insane. Collins had been writing about abnormal and pathological states of mind since the early 1850s, and almost all his fiction is concerned to some degree with medical theory and practice.[17] It was through Dickens, however, that he made most of his

---

[17] See esp. Barbara Fass Leavy, 'Wilkie Collins's Cinderella: The History of Psychology and *The Woman in White*', *Dickens Studies Annual*, 10 (1982), 91–141; and Jenny Bourne Taylor, *In the Secret Theatre of Home: Wilkie Collins, Sensation Narrative, and Nineteenth-Century Psychology* (London, 1988). It is unclear whether Collins was responsible for any of the numerous *Household Words* articles

personal contacts with the institutional management of the mad. On several occasions during the 1850s Dickens professed his ardent admiration and support for the asylum reform movement, of which he took his friend John Conolly to be the guiding figure.[18] Other members of Dickens's circle who had direct private or professional connections with psychological medicine included Thackeray; Charles Reade, active in the Alleged Lunatics' Friend Society, continually firing off letters to the press on the subject and soon to publish *Hard Cash* in *All the Year Round*; John Forster, secretary to the Commissioners in Lunacy from 1853 to 1861, when he became a Commissioner himself; Richard Monckton Milnes, also a Commissioner in Lunacy; Robert Bell, co-proprietor of an asylum at Chiswick; and Bryan Procter.[19] Procter became a Lunacy Commissioner in 1832 and retained the post until 1861 when he retired. Although Thackeray did not follow his friend's advice with regard to Isabella, he was grateful enough to make him the dedicatee of *Vanity Fair*. Procter was also the dedicatee of *The Woman in White*—which suggests that he, rather than John Forster, was probably the source of much of Wilkie Collins's knowledge of contemporary asylum conditions and laws.[20] Neither Collins's nor Dickens's response to the issue of wrongful incarceration can be fully understood, however, without reference to one further person. It was through the production of Edward Bulwer-Lytton's play *Not So Bad as We Seem* in 1851 that Dickens and Collins first met, and

on the subject of insanity, though he certainly collaborated with Dickens on 'The Lazy Tour of Two Idle Apprentices', *Household Words*, Oct. 1857, 313–19, 334–49, 361–7, 385–93, 409–16, which gives a fictional account of their visit to a county lunatic asylum.

[18] See Dickens's speech to the Royal Hospital for Incurables in May 1857, in K. J. Fielding (ed.), *Speeches of Charles Dickens* (Oxford, 1960), 235. Also the collaborative article with W. H. Wills, 'A Curious Dance Round a Curious Tree', *Household Words*, 17 Jan. 1852, 358–9; and, for a list of the main articles on insanity in *Household Words*, Taylor, *In the Secret Theatre of Home*, 250. Dickens's library contained a presentation copy of Conolly's *Croonian Lectures* (1849), inscribed 'with all regard from the Author'. Recorded in *Catalogue of the Library of Charles Dickens from Gadshill: Reprinted from Sotheran's 'Price Current of Literature' Nos. CLXXIX and CLXXV*, ed. J. H. Stonehouse (London, 1935), 23.

[19] Dickens was also an acquaintance of the Earl of Shaftesbury, who was largely responsible for the terms of the 1845 Lunacy Act. See Leavy, 'Wilkie Collins's Cinderella', 109.

[20] See Leigh Hunt, letter to Bryan Procter responding in detail to Procter's account of his visits to asylums on behalf of the Lunacy Commission that year, and commenting favourably on *The Woman in White*; MS letter, 22–3 June 1857, University of Iowa Library (NP 0596785).

it was through him that Dickens in particular became most closely implicated in the lunacy debate.

Lady Lytton's scandalous attack on her husband in her novel *Cheveley* (1839) (see Chapter 5 above) was not the end of public hostilities between this spectacularly ill-matched couple. During the 1840s relations between the Lyttons deteriorated further. Infuriated by her husband's refusal to allow her more than £400 a year, Rosina went out of her way to humiliate him. More vindictive *romans-à-clef* followed, and the press was inundated with letters from her accusing Bulwer-Lytton of brutality, meanness, deceit, adultery, and general viciousness. Several of his friends also received abusive missives, including Dickens and Forster.[21] Rosina wrote up to twenty letters a day (in duplicate) with 'obscene or scurrilous inscriptions', and sent them to her husband's clubs, to the House of Commons, hotels, private homes—anywhere they were likely to be seen.[22] Her grandson likened the effect of reading them years later to 'opening a drawer full of dead wasps. Their venom is now powerless to hurt, but they still produce a shudder and feeling of disgust.'[23]

Rosina's anger might have spent itself had it not been for the death of her 19-year-old daughter, Emily, from typhoid in 1848. Inexplicably, Bulwer-Lytton allowed his child to die in a London boarding-house, and did his best to deny her frantic mother admittance to the bedside. Rosina never forgave him. She blamed him for Emily's death and he blamed her, with the result that her determination to exact public revenge became even more obsessive. Hearing that Dickens and a cast of literary and artistic names including Collins, John Forster, and Robert Bell[24] were to perform *Not So Bad as We Seem* at the home of the Duke of Devonshire, in aid of a new Guild of Art and Literature, she bombarded the Duke with letters denouncing 'Sir Liar' her husband, the 'arch-humbug Dickens', and the whole obnoxious 'Guilt'. She threatened to turn up dressed as Nell Gwyn on the first night, when both the Queen and Prince Albert would be

---

[21] See Victor Alexander, 2nd Earl of Lytton, *The Life of Edward Bulwer, First Lord Lytton*, ii. 267.

[22] Ibid. ii. 266.

[23] Ibid. ii. 267.

[24] For the full cast list, see Edward Bulwer-Lytton, Bart., *Not So Bad as We Seem; or, Many Sides to a Character* (London, 1851), 1.

present,[25] and pelt Victoria with rotten eggs, after which she would expose Her Majesty to the assembled audience as the murderess of Lady Flora Hastings.[26] In the event, she stayed away.

Mr Brooke's unhappy experience at the hustings in Middlemarch was nothing compared to what Bulwer-Lytton went through while celebrating his re-election as the Tory Member of Parliament for Hertford in June 1858 when Rosina finally exacted her revenge. He was thanking the townsfolk for their confidence in him when Rosina made her way on to the platform, where she turned to address the electors. The tirade of accusations included charges that Bulwer-Lytton had murdered her daughter and attempted to murder Rosina herself—claims which she repeated in her autobiography, and which the crowd apparently lapped up.[27] The baronet ignominiously fled the hustings to the jeers and boos of the crowd, and (if Rosina is to be believed) trampled through the flower-beds of a local dignitary's property, leapt over the palings, and locked himself into the house.[28]

Bulwer-Lytton could stand such public humiliations no longer, and sought the professional help of his medical friends. Three weeks later, Rosina was forcibly abducted and committed to Inverness Lodge—a private lunatic asylum in Brentford, run by the renowned alienist Robert Gardiner Hill. It was a disastrous miscalculation—'an act of supreme un-wisdom', as Bulwer-Lytton's grandson put it.[29] Rosina was evidently an immense embarrassment, but there is little to suggest that she was any less sane than her adored mentor Lady Caroline Lamb. The two doctors who

---

[25] The costume may have been a jab at another member of the cast, Douglas Jerrold, whose comedy *Nell Gwynne* had played at Covent Garden in the early 1830s. Rosina may additionally have been aware that Peter Cunningham, also acting in *Not So Bad as We Seem*, was currently writing *The Story of Nell Gwyn; and, The Sayings of Charles II* (London, 1852). Both Jerrold and Cunningham use the story of Nell Gwyn's teasing allusion to Mrs Davis's success in attracting the King by her singing of the mad song from Davenant's *The Rivals* (see p. 11 above); Jerrold, *Nell Gwynne; or, The Prologue, A Comedy, in Two Acts* (London, 1833), 4, 16; Cunningham, *The Story of Nell Gwyn*, 58–60. Rosina's principal intention, however, was clearly to mock her husband's desire to attract the attention of the monarch.

[26] Lytton, *Life of Edward Bulwer*, ii. 267.

[27] Rosina Lytton, *A Blighted Life: A True Story* (London, 1880), 28.

[28] For a fuller account of the Lytton affair, see Marie Mulvey Roberts, Introduction to Rosina Bulwer-Lytton, *A Blighted Life: A True Story* (Bristol, 1994), pp. vii–xxxiv; and John Sutherland, 'Dickens, Reade, *Hard Cash* and Maniac Wives', in his *Victorian Fiction: Writers, Publishers, Readers* (London, 1995).

[29] Lytton, *Life of Edward Bulwer*, ii. 273.

signed the statement recommending her committal were John Conolly and Forbes Winslow (then President of the Medico-Psychological Association).[30] Bulwer-Lytton was probably hoping that their status would prevent an outcry, but he misjudged the power of Rosina's friends and his own political opponents. Outraged letters appeared in the *Daily Telegraph*, the *Somerset County Gazette*, and the *Hertfordshire Gazette* demanding an inquiry,[31] and he was forced to take immediate action. Conolly and Forbes Winslow were obliged to 'revise' their diagnosis. According to Rosina, Bryan Procter was in attendance at her second examination. He was the only one of the Dickens 'clique' she did not despise. In an open letter to Charles Reade, she stated her conviction that Robert Bell had been active for some time in spreading rumours that she was insane.[32] John Forster was accused of being Bulwer-Lytton's 'known tool and toady', setting the rest of the 'Infamous Gang' to attack her reputation and prevent her from publishing.[33] She had no proof that Dickens was directly involved in her abduction and incarceration (and there is nothing in his letters to suggest that he was),[34] but she saw his influence everywhere, and lost no opportunity to smear 'that patent humbug'.[35] After her release from Inverness Lodge, Rosina was taken abroad by her son but he could not manage his travelling companion for long, and she returned to England later that year. She was still attacking 'Sir Liar' long after his death in 1873.[36]

[30] Forbes Winslow was also the founder and editor of the *Journal of Psychological Medicine* and proprietor of two private lunatic asylums. See Richard Hunter and Ida Macalpine, *Three Hundred Years of Psychiatry* (London, 1963), 965, 1074. Conolly was Bulwer-Lytton's representative, while Forbes Winslow acted for Lady Lytton's lawyers. Their certificates are subjoined to a letter written by the Lytton's son Robert to the *Daily Telegraph* to quash rumours that Rosina had been forcibly placed in a lunatic asylum—19 July 1858, 5. See also Louisa Devey, *Life of Rosina, Lady Lytton, with Numerous Extracts from her Autobiography and Other Original Documents. Published in Vindication of her Memory* (London, 1887), 321–4.

[31] Lytton, *Life of Edward Bulwer*, ii. 274. Several of the letters are reprinted in the Supplemental Notes to *A Blighted Life*, 79–95, 97–102.

[32] Lytton, *A Blighted Life*, 12. The book was originally written in response to Reade's appeal for information relating to known cases of wrongful incarceration, printed at the end of *Hard Cash*.

[33] Lytton, *A Blighted Life*, 103.

[34] No correspondence between Dickens and Bulwer-Lytton survives from mid-1858.

[35] Lytton, *A Blighted Life*, 4; and Sutherland, 'Dickens, Reade, *Hard Cash* and Maniac Wives', 72.

[36] James L. Campbell, Sr., *Edward Bulwer-Lytton* (Boston, 1986), 18. Rosina died in 1882.

Dramatic though it was, the Lytton scandal was not the only case likely to have affected Collins's and Dickens's thinking about female insanity in 1859–60. The question of incarcerating 'mad' wives was brought to their attention in other ways. Thackeray's conduct towards Isabella has been strongly defended by his principal biographer, Gordon Ray, who portrays him as a compassionate husband whose response to the tragedy of his wife's madness was unimpeachable. John Sutherland, however, has put a less rosy construction on his behaviour. The decision to place Isabella well out of the way from prying eyes—and in fairly basic conditions to judge from what it cost Thackeray[37]—cannot have been calculated to assist a long-term improvement in her condition. (That it was possible so successfully to remove a deranged wife from the public eye, and from one's own daily life, was probably a strong encouragement to Bulwer-Lytton in 1858.) More disturbingly, Sutherland argues that Dickens himself may have considered employing the threat of the asylum against his own wife. From May 1858, as the Lytton affair was brewing, it was becoming publicly apparent that Dickens's marriage had also broken down. Catherine Dickens was painfully aware of her husband's growing attachment to the 19-year-old actress Ellen Ternan, and she and her mother seem to have made some effort to stop it. Dickens was convinced that his mother-in-law was responsible for rumours that he had committed incest with his sister-in-law, Georgina Hogarth. Discussing the 'violated letter' to his manager[38] in which Dickens declared Catherine unfit for her duties as a wife and a mother, Sutherland draws attention to a phrase which most biographers have found inexplicable: 'her always increasing estrangement' from her children is, Dickens claims, making worse 'a mental disorder under which she sometimes labours'.[39] There is no other evidence that Catherine Dickens was psychologically unstable. Sutherland concedes that it was 'extremely unlikely' that Dickens intended to put his wife in an asylum but 'he may have intended . . . to "show her the instruments". Give her and her mother a terrifying glimpse, that is, of what he *might* do' if they stood between him and Ellen

[37] Sutherland, 'Dickens, Reade, *Hard Cash* and Maniac Wives', 66; but see D. J. Taylor, 'Going into the World', *Times Literary Supplement*, 21 July 1995, 3–4.

[38] The letter was leaked to the *New York Herald Tribune*, which published it without permission. See Walter Dexter (ed.), *The Letters of Charles Dickens*, 3 vols. (London, 1938), iii. 22–3.

[39] Dexter (ed.), *The Letters of Charles Dickens*, iii. 22.

Ternan, or if they failed to assist him in putting down the incest rumours.[40]

The Woman in White (1859–60) and Great Expectations (1860–1) are very much of that context. Written at the height of the lunacy panic, and serialized in close succession in Dickens's journal All the Year Round, both novels have given rise to an extensive secondary literature searching out possible origins for their women in white. Collins's Laura Fairlie and Anne Catherick have been identified with his mistress, Caroline Graves;[41] with Mme de Douhault, the celebrated victim of a wrongful incarceration case in France, whose story was recorded in Maurice Méjan's Recueil des causes célèbres; with the author of a letter that Collins claimed to have received in the late 1850s;[42] and with various other victims of the nineteenth century's asylum laws.[43] Dickens's Miss Havisham has been linked with more real-life recluses than can briefly be listed.[44] As far as à-clef interpretations go, Lady Lytton is clearly closer to Collins's Lady Glyde, consigned to a madhouse by her baronet husband, than to Dickens's Miss Havisham, though Rosina's highly melodramatic response to betrayal, and her determination to exact retribution, may well have shaped Dickens's thinking about abandoned women, insanity, and revenge.

When Charles Reade came to write Hard Cash two years later, he did not hesitate to use his novel to attack criminal irresponsibility in the medical profession.[45] In an afterword, he made a direct appeal to the reader for action over wrongful incarceration:

[40] Sutherland, 'Dickens, Reade, Hard Cash and Maniac Wives', 80.

[41] See Gladys Storey, Dickens and Daughter (London, 1939), 213; quoted in The Woman in White, 604–5. See also Catherine Peters, The King of Inventors: Wilkie Collins (London, 1991), 190–2, 219–20.

[42] Collins, letter to The World, 26 Dec. 1877; repr. as 'Wilkie Collins in Glouces-ter Place', in Edmund Yates (ed.), Celebrities at Home, 3rd ser. (London, 1879); and as app. C of The Woman in White, 588–98.

[43] See Leavy, 'Wilkie Collins's Cinderella', esp. 96–106.

[44] For a summary, see Harry Stone, Dickens and the Invisible World: Fairy Tales, Fantasy, and Novel-Making (Bloomington, Ind., 1979), 279–97. The literature on Miss Havisham's origins is often highly improbable. Particularly far-fetched ac-counts can be found in John Plummer, 'The Original Miss Havisham', The Dicken-sian, 2 (1906), 298 and J. S. Ryan, 'A Possible Australian Source for Miss Havisham', Australian Literary Studies, 1 (1963), 134–6.

[45] See Malcolm Elwin, Charles Reade: A Biography (London, 1931) and Winifred Hughes, The Maniac in the Cellar: Sensation Novels of the 1860s (Princeton, 1980), 86–7.

NOTICE.

I request all those persons in various ranks of life,—who by letter or *vivâ voce* have during the last five years told me of sane persons incarcerated or detained in private asylums, and of other abuses—to communicate with me by letter. I also invite fresh communications: and desire it to be known that this great question did not begin with me in the pages of a novel; neither shall it end there; for, where Justice and Humanity are both concerned, there—

Dict sans faict

À Dieu deplait[46]

Recognizing that sensationalism was necessary to sustain the reader's interest, Reade nevertheless based his novel's claim to merit on the fact that it had 'made something happen', bringing to light numerous cases in which the grounds for committal to an asylum were in doubt (among the letters which he received in response to the appeal was an exhaustive account of her trials by Lady Lytton). Vulnerable to implication in the Lytton affair, neither Collins nor Dickens was so quick to claim superiority over the mad-doctors.

From being an active defender of medical psychology in the 1850s, Dickens was plainly backtracking by the early 1860s. He felt obliged to distance himself from Reade's virulent attack on Conolly in *Hard Cash*, publishing an unprecedented editorial disclaimer after the last instalment in *All the Year Round*, but it is significant that he did not intervene at an earlier stage.[47] Already, in 1862, he had published an article in *All the Year Round*, entitled 'M.D. and MAD', which expressly attacked the authority of medical experts to determine who was or was not certifiable: 'In questions that concern the mind, the less heed we pay to the theorist, and the more distinctly we require none but the sort of evidence patent to the natural sense of ordinary men . . . the better it will be for us. Let us account no man a lunatic whom it requires a

---

[46] Charles Reade, *Hard Cash: A Matter-of-Fact Romance*, 3 vols. (London, 1863), iii. 369.

[47] *All the Year Round*, 26 Dec. 1863, 419. Part of Dickens's motivation was presumably embarrassment at seeing Conolly so viciously satirized as the monomaniacal Dr Wycherley. See Richard Hunter and Ida Macalpine, 'Dickens and Conolly: An Embarrassed Editor's Disclaimer', *Times Literary: Supplement*, 11 Aug. 1961, 534–5; Elaine Showalter, *The Female Malady: Women, Madness, and English Culture 1830–1980* (London, 1987), 48; and Taylor, *In the Secret Theatre of Home*, 41.

mad-doctor to prove insane.'[48] *The Woman in White* and *Great Expectations* precede 'M.D. and MAD', but in different ways they share its distrust of the mid-Victorian alienists' authority, complicating that distrust all the while with an awareness of how far literary men had accepted the thinking behind modern medicine.

Both Collins and Dickens found a means of articulating their ambivalence about the relations between medicine and fiction by returning to the figure of the love-mad woman. Their women in white are discernibly related to the madwomen of sentimental fiction, but they carry the symptoms of love-madness into a distinctly unsentimental world. Like the women in Conolly's, Maudsley's, and Morison's case-studies (see Chapter 2), they inhabit a society in which disappointed affection is subordinated to evidence of physiological disturbance. These women in white fit the outward signs of love-madness, but they no longer make sense to the world, or to themselves, as figures for benignly pitying contemplation. Manipulability is now their dominant characteristic; plots of elaborate complexity are woven around them, and their suffering is often brutally subordinated to the demands of those plots. For Collins, an awareness of the madwoman's disturbing capacity to absorb the meanings attached to her assists a powerful critique of mid-Victorian psychological medicine and a no less exacting critique of romance fiction. Dickens's woman in white stands in a more oblique, but equally sceptical, relation to contemporary theories of insanity and to novelistic convention. Miss Havisham eschews doctors, shutting herself away from the world to grow old and die amid the rotting remains of her wedding-day. She adopts the role of the love-mad woman with such deliberate vengefulness that Estella's question 'Why should I call you mad?' must also be the reader's. *Great Expectations* imagines for its love-mad woman a violent and punitive death; it condemns her with all the moral vigour its narrator can muster; yet it also returns to her an intense desolation, a bitter carelessness of anything but the fact of betrayal, which for the best part of a century the novel had not allowed itself to imagine. Initially denying, but finally reaffirming, the madwoman's capacity to sustain pathos, Dickens's novel forms a troubled reply to the more lightly worn scepticism of *The Woman in*

[48] 'M.D. and MAD', *All the Year Round*, 22 Feb. 1862, 510–13. Cited in Taylor, *In the Secret Theatre of Home*, 103.

White and provides a natural conclusion for the argument which this book has been pursuing.

The reclusive and misanthropic Mr Fairlie in *The Woman in White* has no taste for the sentimental:

Except when the refining process of Art judiciously removes them from all resemblance to Nature, I distinctly object to tears. Tears are scientifically described as a Secretion. I can understand that a secretion may be healthy or unhealthy, but I cannot see the interest of a secretion from a sentimental point of view. Perhaps, my own secretions being all wrong together, I am a little prejudiced as a subject.   (312)

Fairlie's opinions are not endorsed by the novel. Morbidly sensitive, he lives secluded in his darkened rooms at Limmeridge House, admiring his collection of Renaissance prints, 'coquetting' with his coins (318), and exercising, behind his repudiation of sentiment, the refined tyranny of the hypochondriac. He is the primary example of the physical and moral degeneration of the upper classes in *The Woman in White*—and yet his anti-sentimentalism is fully justified by the narrative which unfolds, and his affected repudiation of tears, 'prejudiced' though it certainly is, is consistent with a more widespread desire among those around him to stave off the uncontrolled expression of feeling. Sensationalism and sentimentalism are not usually discussed as related phenomena, yet, as the words themselves might suggest, they are clearly related. The mid-nineteenth-century vogue for 'sensations' repeated many of the forms and much of the logic of eighteenth-century sensibility, but it did so in accordance with a more sceptical vision of society. Where sentimentalism had articulated a paternalistic philosophy—one in which the physiological manifestations of sympathy provided a 'natural' cue to charity—the physical shock that accompanied the mid-Victorian sensation was more likely to require the intervention of the law.

Sensation fiction's recasting of sentimentalism in a harsher mould is clearly set out in *The Woman in White*'s treatment of one of the classic moments in the literature of sensibility. In Sterne's *A Sentimental Journey through France and Italy* (1768), Yorick retravels the road to Moulines earlier taken by Tristram Shandy, and finds the tender spectacle of love-madness he has been promised:

Why does my pulse beat languid as I write this? . . . I discovered poor Maria under a poplar—she was sitting with her elbow in her lap, and her head leaning on one side within her hand—a small brook ran at the foot of the tree. . . . She was dress'd in white, and much as my friend described her, except that her hair hung loose. . . . affliction had touch'd her looks with something that was scarce earthly—still she was feminine—and so much was there about her of all that the heart wishes, or the eye looks for in woman . . .[49]

In Walter Hartright's comparable encounter with the woman in white, pathos has given way to fear:

every drop of blood in my body was brought to a stop by the touch of a hand laid lightly and suddenly on my shoulder. There, in the middle of the broad, bright high-road—there, as if it had that moment sprung out of the earth or dropped from the heaven—stood the figure of a solitary woman, dressed from head to foot in white garments . . . What sort of woman she was, and how she came to be out alone in the high-road, an hour after midnight, I altogether failed to guess. . . . she came close to me, and laid her hand with a sudden gentle stealthiness, on my bosom—a thin hand; a cold hand (when I removed it with mine) even on that sultry night. Remember that I was young; remember that the hand which touched me was a woman's. (15, 17)

Up to a point, the same narrative outline, the same iconography, even the same vocabulary are at work: both scenes involve an impressionable male viewer and a mentally afflicted female subject dressed in white, with an oddness about her situation and her manner that are sexually suggestive. But what was an occasion for pleasurable physical disturbance in *A Sentimental Journey*, causing Yorick's pulse to beat with luxurious languor, has become a life-threatening shock in *The Woman in White*, bringing Hartright's pulse 'to a stop'.

The difference between these two emotionally charged encounters with feminine suffering is plainly not that between 'authentic' sentiment and 'inauthentic' sensation (Sterne was, after all, the most evasive and the most teasing of sentimentalists), but there is, nevertheless, a distinction to be made in terms of their trustworthiness. Sterne offers a sharply observed tableau in which the picturesque pathos of Maria's appearance is not in doubt, though the exact nature of Yorick's response to it is; Walter Hartright, on the

[49] Sterne, *A Sentimental Journey through France and Italy by Mr Yorick*, ed. Ian Jack (London, 1968), 114–16.

other hand, is radically unsure both of what he has seen and of how
to react to it. His meeting with Anne Catherick occurs at night, at
a crossroads on the way from Hampshire to London, and his
powers of observation are steadily undermined: the scene begins in
'mysterious light', it ends in 'black shadows' (14, 21), and
Hartright gives the change deeper significance when he adds that he
traces the words of his story 'distrustfully, with the shadows of
after-events darkening the very paper I write on' (17). Though he
gathers numerous details of Anne Catherick's appearance, he *feels*
he is gleaning almost nothing: 'All I could discern distinctly by the
moonlight . . .'; 'This was all that I could observe.' Sterne's love-
mad woman had 'so much . . . about her of all that the heart wishes
or the eye looks for in a woman'. By contrast, even after he has
conversed with Anne Catherick, Hartright remains radically uncertain of 'What sort of woman she was'.

The shift from the charming tableau of 'Maria at Moulines',
knowingly sought out and pleasurably anticipated by Yorick, to the
unexpected shock of Anne Catherick's touch indicates a profound
change in attitude to physiology. Yorick's meeting with Maria, like
all his sentimental vignettes, leans heavily on contemporary medical
theories of the nervous system. The eye which traces the signs of
affliction in Maria's face is simultaneously the eye of a connoisseur
of feeling and the eye of an amateur eighteenth-century physician.
Indeed, Yorick's whole body is a finely tuned exemplum of the
bodily effects of sympathy—this is a man whose heart beats more
sluggishly at the mere bidding of memory. Walter Hartright's relationship to mid-Victorian medicine is far less easy to determine.
Physiological description enters this scene under strain, so that the
effects of shock on Hartright's frame defy a literal reading: 'every
drop of blood in my body was brought to a stop'.

Collins's hero is, of course, no physician—indeed, it will be one
of Count Fosco's sneers against his antagonist that this upright
Englishman has insufficient understanding of modern medicine.
Fosco claims to have spent the 'best years' of his life 'in the ardent
study of medical and chemical science', and he provides the novel's
most fulsome paean to its power: 'Give me—Fosco—chemistry;
and when Shakespeare has conceived Hamlet, and sits down to
execute the conception—with a few grains of powder dropped into
his daily food, I will reduce his mind, by the action of his body, till
his pen pours out the most abject drivel that ever degraded paper'

(560). Fortunately for society, the mass of modern chemists are not villainous Italian counts but 'worthy fathers of families who keep shops', and—less encouragingly—the average man or woman has a great deal more to fear from the incompetence of medical practitioners than from their malice.

Certainly, the medical profession comes out of the novel badly. Hartright's encounter with Anne Catherick is only the first of numerous occasions on which Collins's characters have cause to doubt, resist, or flatly dispute the judgement of qualified medical men—and it is not only cases of madness which prompt such distrust. Mr Dawson, the respectable family doctor, dismisses Count Fosco as a 'Quack with a handle to his name', yet it is the 'Quack' who correctly diagnoses typhus fever (334–41). When a surgeon does thwart Fosco's scheming it is not through the superior knowledge testified to in the string of letters after his name, '*M.R.C. S. Eng. L.S.A.*', but through his mundane efficiency in registering Anne Catherick's death before the Count can intervene.

Unlike Charles Reade, however, Collins was not simply or even primarily concerned with attacking the medical establishment. Instead, *The Woman in White* greatly complicates its status as a critique of contemporary psychology by questioning how far medical representation and artistic representation remain comparable, and even complicit, in their assumptions and procedures despite the growing institutional power of medicine. Perhaps because Collins had witnessed doctors and novelists colluding in locking away the wife of a baronet only a year before, his novel does not approach the subject of wrongful incarceration with a belief in the moral superiority of literary men. Instead, he draws a series of disturbing parallels between Fosco's medically assisted crime and the aesthetic criteria and the narrative procedures informing Walter Hartright's struggle against medicine.

Walter Hartright is the editor of the several narratives which make up *The Woman in White*. He is also the novel's romantic hero, taking on himself the responsibility for restoring the woman he loves to her rightful place in society, and exonerating Marian from the charge that she is seeking to profit by supporting a madwoman's claim to Lady Glyde's name and property. In all this, he stands on the side of individual identity, against the categorizing force of medical diagnosis. The eyes of a lover, it seems, are subtler than the eyes of a physician. Yet, even as Hartright fights the damage done by an insufficiently discerning medical profession, his language of love repeats the terms of medical error. His first description of Laura Fairlie oscillates strangely between asserting the absolute individuality of her appearance, and blurring her features into soft focus. Hers is a charm 'most gently and yet most distinctly expressed'. Carefully delineated in some respects, she nevertheless continually loses definition: her hair 'nearly melts' into her hat, the beauty of her eyes 'cover[s] and transform[s]' the 'little natural blemishes elsewhere', her face is 'too delicately refined away towards the chin' (41). When he appeals to his readers to supply the image of this woman out of their hearts, Hartright licenses their imaginative freedom: 'Think of her, as you thought of the first woman who quickened the pulses within you that the rest of her sex had no art to stir' (42). But the gesture paradoxically occludes the marks of individuation which it celebrates: it makes an unknown number of faces substitutable for that of Laura. As a blank to be filled with whatever image the reader will supply, she is deprived of the specificity which would have defined not only personality but the difference between sanity and insanity for a Victorian reader: the phrenological bumps that conveyed Rochester's character to Jane Eyre, the cut of the jaw and strength of the brow that conferred individuality. Ostensibly opposed to the medical profession, Hartright's narrative subtly matches the terms of medical knowledge and medical error.

Admittedly, Collins's engagement with Victorian medicine is here, as elsewhere, remarkably indirect. Jenny Taylor has rightly emphasized the degree to which Collins's novel picks up the language and the iconography of the paternalistic psychology of the 1850s, replicating its emphasis on the moral management of the insane,[50] yet the novel offers no direct assessment of contemporary medical theory or practice. The most we see of the mad-doctors responsible for the error on which the novel's plot hangs are a shadowy figure in a carriage and a civil but strange inquisitor whom Laura only half-remembers. Collins says almost nothing about the workings of the asylum,[51] or about its inhabitants. Referred to only as a nameless collective, 'women who were all unknown to her' (393), Laura's fellow patients are never individu-

---

[50] Taylor, *In the Secret Theatre of Home*, esp. 100–3.
[51] D. A. Miller, '*Cage aux folles*: Sensation and Gender in Wilkie Collins's *The Woman in White*', *Representations*, 14 (1986), 107–36: 113.

alized—an omission which puts *The Woman in White* markedly out of line with earlier fictional accounts of the asylum, like those in Maturin's *Melmoth the Wanderer* (1820) or Samuel Warren's *Ten Thousand a Year* (1840), where considerable space was devoted to describing the horrific companions awaiting the victim of wrongful incarceration.[52]

The ease with which Laura's name can be taken from her is only one example of a persistent threat to identity in *The Woman in White*, and that larger context is necessary to an understanding of Collins's anti-sentimental recasting of the love-madness convention. Laura inhabits a world in which unhappiness is indistinguishable from pathology, and in which not only women but men, too, are liable to be reduced to the same set of physiological symptoms. More alarming, because more pronounced, than the theft of her identity is the erosion of Marian Halcombe's individuality. Every critic points to the *doppelgänger* motif in the novel, but the doubling of Laura and Anne is, more properly, a tripling.[53] At the start of the novel, Marian possesses one of the most distinctive faces in Victorian fiction. Hartright's first sight of her provides a finely sensational moment, as the perfect female silhouette turns to reveal a masculine face: swarthy, strong-jawed, with a heavy shadow on the upper lip (25).[54] Yet almost as remarkable as this description is Collins's refusal to exploit it thereafter. The swarthy complexion, 'the dark down' that is 'almost a moustache', and the 'thick, coal-black hair growing unusually low down on her forehead' are never mentioned again. The explanation might simply be editorial tact on Hartright's part: once he comes to know and respect Marian, her 'ugliness' no longer matters. Considered in the context of the plot to deprive another woman of her identity, however, the disappearance of this remarkably individualized face—like the blurred portrait of Laura Fairlie—begins to look more sinister.

Marian provides the most dramatic evidence of the destructive effects of nervousness in *The Woman in White*. When she looks back over her journal, the night before Laura's wedding to Sir Percival Glyde, and notes its escalating hysteria, she cannot

---

[52] For further examples, see Robert S. Surtees's *Handley Cross* (1843); and Henry Cockton's *The Life and Adventures of Valentine Vox, the Ventriloquist* (1844).

[53] Cf. Taylor, *In the Secret Theatre of Home*, 99–107, and 121–2.

[54] See Miller, '*Cage aux folles*', 125–30; and Harvey Peter Sucksmith, Introduction to Wilkie Collins, *The Woman in White* (London, 1975), pp. xviii–xx.

acknowledge the tension as her own: 'Perhaps I may have caught the feverish excitement of Laura's spirits, for the last week' (173).[55] Her impulse is always to assign womanliness to Laura, and to take the masculine part herself, but nervous debility is infectious. D. A. Miller argues that reading itself becomes a hysterical act in the sensation genre[56]—but it is important to add that hysteria also has more specifically literary effects. Reading *The Woman in White*, it is hard not to be affected by the often peculiarly constricted range of its language. 'Fevered' and 'weary', Marian falls into a trance or dream, in which she sees Hartright 'lost in the wilderness', threatened by death repeatedly, and, finally, kneeling by a marble tomb out of which a veiled woman rises to wait by his side. She wakes to see Laura's 'flushed and agitated' face and 'wild bewildered' eyes (249) and to hear her account of meeting the 'wild, frightened', and 'weary'-looking Anne Catherick (255). The same vocabulary echoes between the three women through several pages, implicating them all in the condition and the spectacle of the nervously ill woman. Walter Hartright's first sight of Marian when he returns from America secures the connection. Looking up from Laura's grave he sees her half-sister's face 'Changed, changed as if the years had passed over it! The eyes large and wild, and looking at me with a strange terror in them. The face worn and wasted piteously. Pain and fear and grief written on her as with a brand' (377).

Almost all Collins's characters become prey to similarly dramatic forms of reduction and reduplication—but these are not always nervous in their origins. His interest in doubles is complemented by a fascination with what might be called 'linguistic doubling'. Dictation, rehearsal of speeches, and forgery are all recurrent motifs in *The Woman in White*: Sir Percival obliges Mrs Catherick to pen a confirmation of his good character in his words; Mme Fosco repeatedly takes her cue from the Count and speaks her pre-scripted lines with aplomb; Sir Percival's bastardy is revealed when Hartright discovers that a previous vestry-clerk of Welmingham Church kept a scrupulously accurate copy of the parish registers. On several occasions, the copying of words or of styles is positive, seeming to denote a consciously shared form of language rather than a forcibly imposed one. Following Hartright's direction that their narratives should be told with scrupulous attention to the

---

[55] See also Marian's earlier breakdown (*The Woman in White*, 162–3).
[56] Miller, '*Cage aux folles*'.

truth, relating only what they know from first-hand experience (112), Marian and Mr Gilmore produce personal testimonies remarkably undifferentiated in style. With Hartright, they share a tendency to long lists of adjectives, a fascination with symptoms of mood, and a taste for the jagged narrative style that became the hallmark of sensationalism—the 'interruptive typography' drawing attention to the unpredictability of the narrative flow, and the 'dynamic paragraphing' which accompanies the intrusion of shocking events into the novel's more reflective passages.[57] In an otherwise highly flattering letter, written in response to the first batch of proofs, Dickens offered one criticism:

> I seem to have noticed, here and there, that the great pains you take express themselves a trifle too much ... But on turning to the book again, I find it difficult to take out an instance of this. It rather belongs to your habit of thought and manner of going about the work. Perhaps I express my meaning best when I say that the three people who write the narratives in these proofs have a DISSECTIVE property in common, which is essentially not theirs but yours ...[58]

By 'dissective', Dickens presumably meant to identify the peculiarly forensic quality which Hartright, Marian, and Mr Gilmore all bring to their narratives: the habit of careful analysis of scenes passed through or people encountered; more particularly, the cataloguing of pathological signs in others and in themselves—the fascination with nervousness, with strange moods, and abnormal behaviour. The medical overtones of Dickens's description are highly suggestive. In *The Woman in White*, he suggests, the narrators who pit their skills against medicine nevertheless exhibit a fascination with themselves and others as pathological subjects which invites comparison with medical procedures.

The reduction of people to symptoms in *The Woman in White* is not confined to physical description or to habits of style. Indeed, the novel's slow erosion of the boundaries between literature and medicine is arguably most effective at the larger narrative level, where patterns of behaviour repeat themselves across the novel like symptoms waiting for analysis. Experiences which seem to be characteristic of one person will recur in relation to others: the sight of a

---

[57] On this last point, see John Sutherland, Introduction to Wilkie Collins, *Armadale* (Harmondsworth, 1995), pp. vii–viii.

[58] Dickens, letter to Wilkie Collins, 7 Jan. 1860, in *The Letters of Charles Dickens*, 3 vols., ed. Walter Dexter (London, 1938), iii. 215.

dog trotting at a woman's heels, the putting on of a dark cloak for concealment, the declaration of a passion for Dante, the habit of walking round and round the pond at Blackwater Park. More revealingly, words which have a metaphorical meaning in one context will become suddenly literalized in the experience of another. This subtler form of slippage is most readily recognizable as it affects the term 'asylum'. In the first pages of the novel, Hartright gives the word a metaphorical cast, when he accounts for Professor Pesca's devotion to England: 'he was bound to show his gratitude to the country which had offered him political asylum' (3). Hartright's word for sanctuary, however, will become Laura's actual prison. Later, the term will briefly regain its figurative significance when Fosco promises Marian that Laura's safety will not be threatened if certain rules are kept to ('she has found a new asylum in your heart. Priceless asylum!'; 413)—but Fosco's assurance carries a veiled threat that Laura may at any moment be returned to the madhouse from which Marian rescued her. The damage done by the actual institution has changed everything: the primary meaning of the word, the perception of England, the sense of what a good heart can effect in an unsentimental world.

D. A. Miller notes the fact that the term 'asylum' is twice applied to Limmeridge House late in the novel, and interprets this metaphorical usage as a sign that the methods of social control given institutional force in the madhouse have been internalized into the novel's domestic sphere: the difference between the asylum-as-confinement and the asylum-as-refuge is, he argues, 'sufficiently dramatic to make a properly enclosed domestic circle the object of both desire and—later—gratitude, but evidently it is also sufficiently precarious to warrant—as the means of maintaining it—a domestic self-discipline that must have internalized the institutional control it thereby forestalls'.[59] To emphasize the instability of the term asylum is, however, to risk overlooking what is most disturbing—and plainly so—about *The Woman in White*: its depiction of a world in which private unhappiness is rarely permitted to remain internalized. Psychological oppression takes on remorselessly material form in Collins's novel; metaphor time and again becomes

---

[59] Miller, '*Cage aux folles*', 113. See also Taylor, *In the Secret Theatre of Home*, 99, arguing that the breaking down of Laura's selfhood 'in turn breaks down any stable division between the resonances of "home" and "asylum" as places of safety and danger'.

reality, and the return of Laura from the madhouse to the home is one of the very few occasions on which that movement is reversible. The first meaning of a disputed or unstable term in this novel is almost always the metaphorical one, and a 'sensation' is accomplished by making it literal. So, Fosco is condemned as 'a spy' long before it is recognized that he actually is one (269, 524–5); the 'brand' is a mark of sorrow and pain before Pesca discloses the mark of the secret society 'deeply burnt in the flesh' (377, 537); 'wildness' looks out of many frightened eyes before Hartright faces the real 'wilds of Central America' (249, 253, 373); and Sir Percival has given ample demonstration of his 'smouldering temper' before he is literally consumed in flames (223, 476–85).

The insistent drive towards literalization and the fear that its effects may be irreversible hold the key to *The Woman in White*'s particular interest as a reworking of the love-madness convention. If novels had representative organs *The Woman in White*'s would unquestionably be the heart. Less diffusive than the nerves, the heart is also a more 'old-fashioned' organ, central to sentimentalism but also to romance—and acutely vulnerable in the context of sensationalism. The faltering cardiac beat dominates the story's sensational repertoire, from Hartright's arrested pulse when Anne Catherick touches his shoulder onwards. The 'feeling' heart, the 'softened', 'aching', and 'penetrated' heart, are mainstays of the novel's sentimental and romantic vocabulary, yet its characters increasingly find themselves in a world where the heart can be more directly targeted. Sir Percival Glyde first meets Count Fosco on the steps of the Trinità del Monte in Rome, when a timely intervention by the Italian prevents Sir Percival from being 'wounded in the heart' by robbers (172). Fosco himself will eventually die from a blow 'struck with a knife or dagger exactly over his heart' (581). And the crux of the narrative—the success or failure of Fosco's plot to substitute Laura, Lady Glyde for Anne Catherick— hinges on Hartright's ability to prove that Anne died in London of heart disease before Laura left her home in Blackwater Park, Hampshire.

Anne's pallid face is the most obvious symptom of her illness: the 'serious affection of the heart' which troubles her from the time she hears of Laura's engagement to Sir Percival Glyde (424). The ailment sounds—and, in its first manifestation, is—heavily symbolic. The timing of her illness indicates that Anne suffers on behalf of

Laura, just as Laura suffers from being incarcerated in Anne's name. Recalling the double-plotting of earlier nineteenth-century novels like *Madelina* and *Glenarvon*, *The Woman in White* sets up the less socially advantaged woman to experience her half-sister's sickness and, finally, her death for her, leaving Laura free to return to her lover and to enjoy—eventually—a happy ending. Yet, having established that narrative structure, Collins soon begins to whittle away at it, questioning the meaning of Anne's illness, and disputing the terms which allow her to function as Laura's romantic double. Meeting Anne again in Cumberland, Hartright suspects, not for the first time, that her story might be a 'too common and too customary' one, and he gently interrogates her. When she alludes to her 'misfortune', he asks what she means by the word:

'The misfortune of my being shut up,' she answered, with every appearance of feeling surprised by the question. 'What other misfortune could there be?' . . .

'There is another misfortune,' I said, 'to which a woman may be liable, and by which she may suffer life-long sorrow and shame.'

'What is it?' she asked, eagerly.

'The misfortune of believing too innocently in her own virtue, and in the faith and honour of the man she loves,' I answered.

She looked up at me, with the artless bewilderment of a child. Not the slightest confusion or change of colour; not the faintest trace of any secret consciousness of shame struggling to the surface, appeared in her face—that face which betrayed every emotion with such transparent clearness. (88)

In her 'artless' way, Anne Catherick confuses Hartright's judgement, showing his assumption to be 'plainly and distinctly the wrong one' (88). In doing so she wordlessly rules out the conventional explanation for her presence. Clothed in the familiar white dress of the love-mad woman, declaring herself to have been betrayed by a man, and suffering evident symptoms of mental distress, she is nevertheless an innocent, bewildered by Hartright's questioning, and unable to comprehend his language. Requiring him to relinquish the scenario of the abandoned woman, she leads him gradually to recognize that the woman in front of him might be the victim not of her own insufficient virtue but of a man's more calculated crime. In doing so, she sets up an anti-sentimental framework for understanding Laura which Hartright will claim to have understood but which he will in fact bitterly resist—for to be impervious to the tragic forms of romantic convention is to be impervious also to its consolations.

Laura, like Anne, appears to have the makings of a love-mad woman. She, too, wears the white dress and possesses the naïve simplicity and emotional vulnerability that would lead the reader to expect a Crazy Jane plot. She also meets the essential criterion for love-madness absent from Anne's story: the loss of her lover. Debarred from marrying Walter Hartright, she is compelled to marry Sir Percival Glyde in fulfilment of an 'engagement of honour' sanctioned by her father on his deathbed. But here Laura's similarity to the madwomen of sentimental fiction begins to break down. Her behaviour from the time she knows she must marry Sir Percival indicates a determined effort at self-control. She ritualistically puts away the mementoes of her summer with Hartright, and determines to submit to the 'hard duties' of her new life (154). To be 'inflexibly passive', is, of course, not the same thing as to assert one's will, but once Percival has shown himself a brutal and intolerable husband, Laura does display a more healthy resistance. Yet, the principles which enabled Jane Eyre to maintain her sanity are signally inadequate to save Laura from the asylum. Instead, Count Fosco's success in confining her to the madhouse under Anne Catherick's name involves a brutal rejection of the sufficiency of self-control even where every effort of will is made. The will is not enough in a world where, as Fosco repeatedly reminds us, chemical science can render the most powerful of minds nothing.

In stealing her identity from Laura and giving her that of Anne Catherick instead, Sir Percival Glyde and Count Fosco force upon her a role which she has done everything possible to avoid. They exploit the invisibility of the difference between an emotionally afflicted heart and a physically weakened one, and in doing so create the terms for a critique not only of medicine but of the romantic assumptions which, in the hands of other writers, had provided a secure means for challenging medical expertise. Hartright describes the dangerous pressure of metaphor towards literalism and of romantic illusion towards realism when he examines Laura's face after her time in the asylum: 'The sorrow and suffering which I had once blamed myself for associating even by a passing thought with the future of Laura Fairlie, *had* set their profaning marks on her face; and the fatal resemblance which I had once seen and shuddered at seeing, in idea only, was now a real and living resemblance which asserted itself before my own eyes' (399–400). Heavily overwritten, the passage keeps reiterating its point: sorrow has made a visible mark, fatality has become reality, mere idea has become an insistent fact, 'assert[ing] itself'—there—'before my own eyes'. Psychological suffering has once again taken on the less tractable form of physiological damage.

Hartright's response is to begin immediately restoring the possibility for romance: 'whatever long, weary, heart-breaking delays it involved, the wrong that had been inflicted on [Laura], if mortal means could grapple with it, must be redressed' (401). His language establishes his intentions even before he puts them into practice, returning the 'broken heart' to the realm of metaphor. Through painstaking nursing and gentle encouragement he and Marian do indeed succeed in restoring Laura to health, and *The Woman in White* rewards their efforts. Laura is at last given back her name and her inheritance and Anne Catherick is officially laid to rest. Hartright's credentials as the novel's defender of romance are, of course, inscribed in his name and the fact that his son will unite that name to Laura's wealth seems to confirm the restoration of the romance genre as a viable mode. Given the scepticism this novel has induced about the romantic idiom, however, Hartright's conclusion is always going to be vulnerable to the charge of naïvety. More pointedly, he is not the only one to lay claim to the vindicating power of romance.

'I have to assert', Count Fosco writes at the end of his exhaustive confession, 'that the one weak place in my scheme would never have been found out, if the one weak place in my heart had not been discovered first' (569). A 'fatal admiration' for the 'sublime' Miss Halcombe caused him to be untrue to himself at a critical moment. 'At the ripe age of sixty, I make this unparalleled confession. Youths! I invoke your sympathy. Maidens! I claim your tears' (570). Undercutting Hartright's claim to have allied romance with virtue, Fosco makes sentimentalism the basis of his own appeal for a reprieve from the reader. Knowing as he does that, when their sentiments are engaged, the English are not morally discriminating (212–13), he plays the sentimental card for all it is worth, but the appeal for sympathy concludes on a satiric note. That Fosco should introduce his account of the plot to confine Lady Glyde to a madhouse by echoing Thackeray's conclusion to *Vanity Fair* was probably an unintended irony, but, deliberate or otherwise, it underlines the novel's persistent refusal to see literature as a sphere innocent of the deceptions and the moral complications which were besetting modern medicine: 'Such is the World; such Man; such Love. What are we (I ask) but puppets in a show-box? Oh, omnipotent Destiny, pull our strings gently! Dance us mercifully off our miserable little stage!' (558).

When serialization of *The Woman in White* ended in *All the Year Round* in August 1860, sales dropped off abruptly. Charles Lever's *A Day's Ride: A Life's Romance* was an inadequate substitute, and in order to rescue the circulation figures, Dickens was obliged to begin publishing *Great Expectations* in weekly rather than monthly instalments. There are enough similarities between *Great Expectations* and *The Woman in White* to suggest that Dickens was influenced by his friend's novel and hoped to benefit from its success. He congratulated Collins on having found 'the name of names, and very title of titles' for his novel and, as Harry Stone has shown, this enthusiasm reflected a long-standing fascination with stories of reclusive and maddened women in white.[60] Like Collins, Dickens was interested in the questionable status of late eighteenth-century sentimental values seventy years on. Many set-pieces of sentimental fiction were revitalized in his novels (the lingering deathbed scene, the repentance of the prostitute, the return of the lost child) but the love-mad woman singularly failed to attract his sympathy. In Miss Havisham, Crazy Jane is grotesquely travestied. Most critics have seen this jilted bride as an image of trauma: a woman who has arrested her life at the moment of her betrayal. In fact, she has neither preserved the past nor sustained its pathos. She may have stopped the clocks, but her life is a bitter testament to the power of devouring Time.

Pip's first, childish impression is of a woman frozen in the reflective posture beloved by nineteenth-century portrait painters. Sitting in an armchair, 'with an elbow resting on the table and her head leaning on that hand', she is imposing, but she is also, the adult Pip warns, 'the strangest lady I have ever seen, or shall ever see':

She was dressed in rich materials—satins, and lace, and silks—all of white. Her shoes were white. And she had a long white veil dependent from her hair, and she had bridal flowers in her hair, but her hair was white. Some

---

[60] Stone, *Dickens and the Invisible World*, 279–97.

bright jewels sparkled on her neck and on her hands, and some other jewels lay sparkling on the table. Dresses, less splendid than the dress she wore, and half-packed trunks, were scattered about. She had not quite finished dressing, for she had but one shoe on—the other was on the table near her hand—her veil was but half arranged, her watch and chain were not put on, and some lace for her bosom lay with those trinkets, and with her handkerchief, and gloves, and some flowers, and a prayerbook, all confusedly heaped about the looking-glass.   (56)

This is Pip's first encounter with wealth, and his eye restlessly accumulates: 'and . . . and . . . and'. Initially he is distracted from Miss Havisham herself by the mass of materials and possessions surrounding her: their richness and colour, the brightness of the jewels, and the clutter of objects. The one jolt comes with the awareness of something wrong but not yet dwelt upon: 'but her hair was white'. At first the adjectives are few and mainly impressionistic, but as Pip takes in more of the bridal scene, the qualifications come with a rush. The only part of the description which remains undamaged by revision is the 'white hair'.

It was not in the first few moments that I saw all these things, though I saw that everything within my view which ought to be white, had been white long ago, and had lost its lustre, and was faded and yellow. I saw that the bride within the bridal dress had withered like the dress, and like the flowers, and had no brightness left but the brightness of her sunken eyes. I saw that the dress had been put upon the rounded figure of a young woman and that the figure upon which it hung loose, had shrunk to skin and bone.   (87)

To gaze at Miss Havisham is to experience disillusionment. Pip registers her degradation as a resurgence of his own first acquaintance with the reality of death, and terror induces in him a paralysis not unlike the arrested motion of the scene before him. Panic prevents any attempt to turn away his eyes: 'I saw . . . I saw . . . I saw . . .'. In the act of embodying her own cruellest memory, Miss Havisham triggers the resurgence of *his* past nightmares: a 'ghastly waxwork' once seen at a fair, a 'skeleton in the ashes of a rich dress . . . dug out of the vault under a church pavement' (87). It is a shock to realize, later, that this woman is only in her early forties, for she has about her the quality of extreme old age: the sunken eye, and shrunken body of a death-in-life or life-in-death figure. In Pip's imaginings, she blurs the boundaries between the two states, terrifying him, as the two men in the graveyard terrified him earlier, by presenting life where death should be. He makes the connection himself when he sees her hands on her heart and thinks, immediately, of the young man who Magwitch threatened would tear out his heart and liver to be 'roasted and ate' (6).

Miss Havisham should be tragic, but there is a forced quality to her desolation, even in this first appearance, that leaches sympathy from her. Had Pip told us that 'the bride had faded like the dress, and like the flowers', this would be sentimentalism; but 'withered' is too strong, too literal, for easy sympathy, and the repetition of 'bride . . . bridal' insists too much. As if afraid of being misinterpreted, she puts the boy through a brief catechism, making sure that he understands the meaning of her appearance:

'Look at me,' said Miss Havisham. 'You are not afraid of a woman who has never seen the sun since you were born?'

I regret to state that I was not afraid of telling the enormous lie comprehended in the answer 'No'.

'Do you know what I touch here?' she said, laying her hands, one upon the other, on her left side.

'Yes ma'am.' (It made me think of the young man.)

'What do I touch?'

'Your heart.'

'Broken!'

She uttered the word with an eager look, and with strong emphasis, and with a kind of weird smile that had a kind of boast in it. Afterwards, she kept her hands there for a little while, and slowly took them away as if they were heavy. (57)

The gestures might have been learned straight from an actor's manual; the speech has the mannered quality of stage delivery; the pause seems calculated for effect. Like her later exhibition of distress, rising up in her chair and striking at the air, the performance risks treading on the wrong side of melodrama, and collapsing into bathos.

One of Miss Havisham's closest antecedents was, in fact, a failed theatrical performance. In 1831 the comic actor Charles Mathews introduced into one of his celebrated 'At Homes' a well-known London figure, 'the White Woman', who could be seen haunting Oxford Street at that period. The exploitation of a genuine case of distress did not go down well with the opening-night audience, and the protests forced Mathews to withdraw the sketch.[61] Martin Meisel notes that, as a keen attender of the 'At Homes', it is likely that Dickens saw the piece notwithstanding its prompt removal from the programme. Certainly he was familiar with the original. In January 1853 the White Woman of Oxford Street made her appearance in *Household Words*, amongst a series of vignettes entitled 'Where We Stopped Growing'. The article offered a whimsical meditation on the power of certain strange figures to impress themselves so firmly in our minds in our childhood that their image remains a source of undiminished fascination, not unmixed with a kind of threat. One such figure, for Dickens, was the White Woman. In his mind she exists always in the present tense:

> She is dressed entirely in white, with a ghastly white plaiting round her head and face, inside her white bonnet. She even carries (we hope) a white umbrella. With white boots, we know she picks her way through the winter dirt. She is a conceited old creature, cold and formal in manner, and evidently went simpering mad on personal grounds alone—no doubt because a wealthy Quaker wouldn't marry her. This is her bridal dress. She is always walking up here, on her way to church to marry the false Quaker. We observe in her mincing step and fishy eye that she intends to lead him a sharp life. We stopped growing when we got at the conclusion that the Quaker had had a happy escape of the White Woman.[62]

The White Woman's story would once have been an eagerly grasped occasion for sentimentalism, but it has now become a subject of scorn and of punitive anger. She prompts no overt expression of sympathy from Dickens. The fishy eye repels. When he concludes that 'she evidently went simpering mad on personal grounds alone' Dickens rules out the assumption which had formerly sustained sentimentalism: the unquestioned authenticity of an abandoned woman's grief. This woman has the air of a bad actress. Garishly dressed, over-painted, cold and conceited, she is not pathetic but grotesque. Yet, comedy looks very like a defence against her claim on the viewer. As if in spite of its narrator, there *is* sympathy at work in the description. The white woman has ingrained herself in his memory so intimately that an act of conscious sympathy has become redundant: description leads repeat-

[61] See Martin Meisel, 'Miss Havisham Brought to Book', *PMLA*, 81/3 (1966), 278–85: 278–80.

[62] Dickens, 'Where We Stopped Growing', *Household Words*, 1 Jan. 1853, 361–3.

edly to imaginative embellishment—'she carries (we hope) a white umbrella', 'no doubt because a wealthy old Quaker wouldn't marry her'. Significantly, the moment of psychological arrest is not the sight of her, but the recognition that his sentimental expectations have been violated: 'we stopped growing when we got at the conclusion that the Quaker had had a happy escape'.

Miss Havisham exerts comparable power over Pip's imagination, and his desire to repudiate her at the end of the novel is brutally apparent, yet she seems immune from the kind of wit that Dickens had earlier directed at the White Woman of Oxford Street. George Bernard Shaw found himself tempted to compare Miss Havisham with another, violently misanthropic Dickensian character, Mr F.'s aunt in *Little Dorrit*, and confessed that it seemed 'hardly decent . . . but as contrasted studies of madwomen they make you shudder at the thought of what Dickens might have made of Miss Havisham if he had seen her as a comic personage'.[63] That he resisted seeing her in that light is the more remarkable given the ease with which abandoned women turn into objects of laughter elsewhere in *Great Expectations*.

When Mr Wopsle, the church clerk turned travelling actor, performs *Hamlet* in 'a small metropolitan theatre' his audience has no squeamishness about critical 'decency': 'Whenever that undecided Prince had to ask a question or state a doubt, the public helped him out with it. As for example; on the question whether 'twas nobler in the mind to suffer, some roared yes, and some no, and some inclining to both opinions said "toss up for it;" and quite a Debating Society arose' (251). Among the objects of their derision is Wopsle's Ophelia: 'Ophelia was a prey to such slow musical madness, that when, in course of time, she had taken off her white muslin scarf, folded it up, and buried it, a sulky man who had been long cooling his impatient nose against an iron bar in the front row of the gallery, growled, "Now the baby's put to bed let's have some supper!"' (251). Like Wopsle in his improbable wig, dying 'by inches from the ankles upward' (276), this Ophelia drags the tragedy out beyond all reasonable limits and tips it over into bathos. Instead of wooing her audience with the pathos of her insanity, she provokes a vision of industrious motherhood. As Pip puts it, the suggestion is, 'to say the least of it . . . out of keeping'.

---

[63] Preface to the Edinburgh limited edition of *Great Expectations* (1937), p. vi.

The novel's other bereaved and deserted women are no more successful in sustaining their tragic significance. The lawyer's clerk, Mr Wemmick, invests heavily in 'portable property'—gifts from those clients whom even Mr Jaggers could not save from the gallows. Among the several pieces of mourning jewellery which Wemmick wears about his person is a brooch 'representing a lady and a weeping willow at a tomb with an urn on it' (169). On a visit to the office to request money from Mr Jaggers, Pip learns from Wemmick the story of its origins. Taking down two dusty death masks from the wall of the office, the clerk presents the first as 'Old Artful', a celebrated client, hanged for murdering his master.

'You had a particular fancy for me, hadn't you Old Artful?' said Wemmick. He then explained this affectionate apostrophe, by touching his brooch . . . and saying, 'Had it made for me, express!'
    'Is the lady anybody?' said I.
    'No,' returned Wemmick. 'Only his game. (You liked your bit of game, didn't you?) No; deuce a bit of a lady in the case, Mr Pip, except one—and she wasn't of this slender ladylike sort, and you wouldn't have caught *her* looking after this urn—unless there was something to drink in it.'   (198)

Like the account of Mrs Gargery's absurdly over-managed funeral, Wemmick's mourning-jewellery provides Dickens with an opportunity for reflecting satirically on the expanding death industry of the mid-Victorian period. In associating the 'slender ladylike sort' represented by Miss Havisham and Estella with Old Artful's woman, Wemmick also quietly anticipates one of the sensations of *Great Expectations*. Estella's mother, Molly, also has the slim form, pale face, 'large faded eyes', and 'streaming hair' of the love-mad woman, but, once again, sentimentalism is invoked only to be travestied. Far from pining hopelessly for her lost lover, she has murdered her rival for the man's affections and escaped justice for it. Outside the barred gates of Satis House, abandoned women are debased figures: comic, or criminal, or both.[64]

---

[64] The novel's repeated allusions to George Lillo's *The London Merchant; or, The History of George Barnwell: A Tragedy* (1731) suggest another, more distant, joke at the expense of the love-mad woman. At the end of the play, George Barnwell is condemned to death for embezzling his uncle's money and then murdering him. His cousin, Maria, comes to the jail and, in deep distress, confesses her love for the renegade. As he goes to the gallows, she seems set to descend into love-madness. However, in an epilogue written by Collie Cibber in 1731 and performed by his wife in the character of Maria, she returns on stage and offers her heart to whichever man in the audience will applaud the play most loudly.

Miss Havisham, however, has insinuated herself so deeply into Pip's mind that comedy at her expense would have to become self-parody. When he reflects on the formation of that first link 'in the long chain of iron or gold, of thorns or flowers' that bound him from his first visit to Satis House (71) he offers the reader another buried clue to the link between Magwitch's world and that of Miss Havisham,[65] but he also sums up the gift that Satis House confers on him. Magwitch makes Pip a gentleman, but Miss Havisham makes him a story-teller—a 'liar' in Joe Gargery's blunter terms— teaching him the trick of language and of mind that makes a 'chain of gold' substitutable for a 'chain of iron'. In making him capable of that kind of imaginative transformation, she makes it possible for Magwitch's money to work its magic. When Mr Jaggers announces that Pip is to be 'immediately removed from his present sphere of life and from this place, and be brought up as a gentleman' his words are the fulfilment of a cherished fantasy: 'I had always longed for it' (135, 137).

To say, however, that Miss Havisham teaches Pip the *meaning* of metaphor would be untrue. In fact, she teaches him to resist any distinction between what is literal and what is figurative. When he listens to her, he learns that all the fear he felt in the hands of Magwitch and more is bound up in the meaning of the word love:

'Love her, love her, love her! If she favours you, love her. If she wounds you, love her. If she tears your heart to pieces—and as it gets older and stronger, it will tear deeper—love her, love her, love her!'
Never had I seen such passionate eagerness as was joined to her utterance of these words. I could feel the muscles of the thin arm round my neck, swell with the vehemence that possessed her.
. . . 'I'll tell you,' said she, in the same hurried whisper, 'what real love is. It is blind devotion, unquestioning self-humiliation, utter submission, trust and belief against yourself and against the whole world, giving up your whole heart and soul to the smiter—as I did!' (237)

Translating Magwitch's threat of physical violence ('tear out your heart and liver to be roasted and ate') into a definition of love ('tears your heart to pieces'), Miss Havisham insists on the equivalence of physical and emotional damage. Like Collins's women in

[65] In addition to the obvious link through the word 'chain', Christopher D. Morris has drawn attention to the phrase's likeness to an inscription on a tombstone; see 'The Bad Faith of Pip's Bad Faith: Deconstructing *Great Expectations*', *ELH*, 54 (1987), 941–55.

white, she lives in a world where 'hard facts' are privileged over feeling, physiology over psychology. Like Hartright, she resists that view of the world, but unlike him she does so not by attempting to revalidate romance but by living out the romance scenario with unremitting attention to its physical cost. None of Pip's encounters with Estella is as intimate as this moment when Miss Havisham tightens the thin muscles of her arm around his neck. Clutching him, beating her fist in the air, appearing 'as if she would have struck herself against the wall and fallen dead' (237) she gives her words physical force, belying their status as 'mere' words.

Like Wilkie Collins's women in white, Miss Havisham offers an indirect engagement with medicine, but one which takes much of its power from that indirectness. Her insistence that emotional suffering is as 'real' as physical suffering implies a deep resistance to contemporary psychological medicine's increasing emphasis on physiological explanation. All her energy goes into maintaining the physical evidence of her pain, preserving her own image and the wedding scene around her at just that level of decay where the form remains recognizable: mending the worst tears in the bridal dress, feeding her skeletal frame on what scraps she can lay her hands on in the night. Her story will end only when she has become physically absorbed into this scene of ruination, taking the place of the rotten wedding-cake on the table, consumed like it, but by 'sharper teeth than teeth of mice'. Then, metaphor will have blended into reality, and she will be a safe object for the world's wonder: 'They shall come and look at me here' (83).

To be seen on any other terms would, it seems, be to risk devaluing herself. Christopher Ricks is right to suggest that *Great Expectations* causes peculiar problems for a critic because 'the most important things about [it] are also the most obvious. . . . There needs no explicator to tell us what is primarily great about Miss Havisham and her decaying house.'[66] Yet, it is also the case that few of the characters within the novel seem capable of accepting the obvious. For those few people allowed to see Miss Havisham, she always means something other than her own desolation. Throughout nineteenth-century fiction, love-mad women had been used to facilitate the telling of other stories, and Miss Havisham is no exception. For her cousins, Sarah, Camilla, and Raymond, 'toadies

[66] Christopher Ricks, '*Great Expectations*', in John Gross and Gabriel Pearson (eds.), *Dickens and the Twentieth Century* (London, 1962), 199.

and humbugs' in Pip's eyes, she means a possible inheritance, and is therefore to be humoured and flattered: 'Dear Miss Havisham', enthuses Sarah Pocket, absurdly: 'How well you look!' 'I do not . . . I am yellow skin and bone' is the remorselessly factual reply (84). For Mr Jaggers, she is 'an eccentric old lady' (408), but above all a client. For Estella she is a hard taskmistress, and an intricate plotter of revenge (301). Joe, perhaps, comes closest to appreciating Miss Havisham on the terms she would wish for, when he finds himself reduced to the single adjective: 'Astonishing!'—but even he fabricates a lie about her in order to appease Mrs Joe: 'Miss Havisham . . . made it wery partick'ler that we should give—were it compliments or respects, Pip?' (101)

Pip's interpretation of Miss Havisham is altogether more ambitious. Her disappointed expectations should provide a monitory example to him, but he is incapable of heeding her advice to 'expect no other and no more reward' than the £25 she gives in payment of his indenture as an apprentice to Joe. Instead, he flatly disobeys and interprets his own transition from the world of the blacksmith's forge to the life of a leisured gentleman as evidence that he is the fairy-tale prince, elected by her to break the spell of Satis House.[67] For Pip, as many commentators have remarked, Miss Havisham *is* gentility. He learns her story at the same time that he acquires his first lessons in good manners from Herbert Pocket. The account of her wealthy family background, her spoilt childhood, the disinheritance of her brother, her attraction to Compeyson, his financial exploitation of her, and the ill-fated wedding-day, is interspersed with friendly directions about the correct manner of using a knife and fork, and the amount of food that may politely be placed in the mouth at one time. Herbert's tactful advice to Pip seems to endorse Dickens's liberal belief in social advancement through self-help, but Miss Havisham's example ought to alert Pip to a less tractable perspective on class. Gentility is a peculiarly unchallengeable notion in the Havisham–Pocket family: it may be lost, but it certainly cannot be acquired or recovered, as Mrs Pocket's obsessive attempt to locate the point at which her grandfather ought to have come into his baronetcy suggests. Miss Havisham's father was a gentleman and a brewer: 'I don't know why it should be a crack thing to

---

[67] On *Great Expectations* and fairy-tale, see Shirley Grob, 'Dickens and Some Motifs of the Fairy Tale', *Texas Studies in Literature and Language*, 5 (1965), 567–79, and Stone, *Dickens and the Invisible World*, 298–339.

be a brewer', Herbert Pocket confesses, 'but it is indisputable that while you cannot possibly be genteel and bake, you may be as genteel as you like and brew' (178). Compeyson repeatedly betrayed his lack of gentility: he was 'a showy-man . . . not a true gentleman at heart'. The implications for Pip, endeavouring to manage his knife and fork correctly on the other side of the table, are not encouraging: 'no varnish can hide the grain of the wood; and the more varnish you put on, the more the grain will express itself' (179).[68]

Why does Pip not take warning, as both Miss Havisham and Estella repeatedly urge him to do? Far from rebuking himself for his own wilful misinterpretation, he finds it a short step from recognizing that Miss Havisham was not his benefactress to blaming her for allowing him to persist in his delusion. Having learnt that his money comes not from Satis House but from the penal colony in Australia, and that he is the bought creature of a convict, he charges Miss Havisham with complicity. She is unrepentant:

'Yes,' she returned, again nodding steadily, 'I let you go on.'
'Was that kind?'
'Who am I,' cried Miss Havisham, striking her stick upon the floor and flashing into wrath so suddenly that Estella glanced up at her in surprise, 'who am I, for God's sake, that I should be kind?' (355)

Pip recognizes that it was 'a weak complaint to have made', but he does not forgive. Having learnt his lessons at her side, he acts as if his delusions, as much as his money, were gifted to him by a selfish donor. There is, at least, a kind of reciprocity in his recognition that, if she has taken from him the illusion of romance, he can take from her the illusion of revenge.

Miss Havisham's revenge depends on Pip taking on the role of a new Compeyson, coming to Satis House, courting its heiress, with the difference that he would be spurned, not loved in return. But the scene which unfolds before her eyes as Pip declares his love to an immovable Estella fails to bring her satisfaction. Seeing her desolation now mirrored in Pip's misery, Miss Havisham is not triumphant but vulnerable once again to feeling and therefore to loss.

---

[68] On class in *Great Expectations*, see Julian Moynahan, 'The Hero's Guilt: The Case of *Great Expectations*', *Essays in Criticism*, 10 (1960), 60–79; T. A. Jackson, *Charles Dickens: The Progress of a Radical* (London, 1937), 188–200; and Anny Sandrin, *Great Expectations* (London, 1988), 47–59.

Pip is never closer to Compeyson than when he leaves her, like a departing lover, using the same word that will occur more ambiguously in the revised ending of the novel: 'No matter with what other words we parted; we parted' (396). This leave-taking should suggest pity, closure, even a subtly effective kind of revenge, yet it is followed by a moral of startling harshness:

That she had done a grievous thing in taking an impressionable child to mould into the form that her wild resentment, spurned affection, and wounded pride, found vengeance in, I knew full well. But that, in shutting out the light of day, she had shut out infinitely more; that, in seclusion, she had secluded herself from a thousand natural and healing influences; that, her mind, brooding solitary, had grown diseased, as all minds do and must and will that reverse the appointed order of their Maker; I knew equally well. And could I look upon her without compassion, seeing her punishment in the ruin she was, in her profound unfitness for this earth on which she was placed, in the vanity of sorrow which had become a master mania, like the vanity of penitence, the vanity of remorse, the vanity of unworthiness, and other monstrous vanities that have been curses in this world? (394)

Dickens probably shared Pip's sentiments. Writing to Collins two years earlier, he had offered his friend the outline of an idea for a Christmas story in *All the Year Round*. The story should concern a man or a woman 'prematurely disgusted with the world' who would, like Scrooge, be compelled to recognize their error: 'everything that happens' should show that 'you can't shut out the world; that you are in it, to be of it; that you get into a false position the moment you try to sever yourself from it; and that you must mingle with it, and make the best of it, and make the best of yourself into the bargain'.[69] The same message motivates a great deal of Dickens's writing, most famously in *A Christmas Carol* (1843), but also in numerous other works, including the Mrs Clennam chapters of *Little Dorrit* (1857), and the short story 'Tom Tiddler's Ground' (1861).[70] As a summary of Miss Havisham's life, however, Pip's words insist too much: the falling cadences, the elaborate parallels, the talismanic 'I knew . . . I knew' are not elegy but self-vindication. Once again, Miss Havisham seems to mean more than she should;

[69] Letter to Collins, 6 Sept. 1858, in *Letters of Charles Dickens*, iii. 50–1.
[70] The extra Christmas number of *All the Year Round*, Dec. 1861. See also the ballad 'The Dirty Old Man: A Lay of Leadenhall', *Household Words*, 8 Jan. 1853, 396–7, possibly of Dickens's composition.

indeed, in this instance, Pip's words betray a kind of desperate attempt to ensure that she has a secure meaning for him. It is a curious feature of *Great Expectations* criticism that so much of it feels obliged to confirm the justice of Pip's judgement on her, for the novel itself is less content with this attempt at closure. Far from resolving his relationship to Satis House, the hard moral proves the prelude to a scene of more disturbing vengefulness.

Like several other scenes in *Great Expectations*—Pumblechook's drinking of the tar-water, Joe's thrashing of Orlick, and, more particularly, the attack on Mrs Joe, and the later attack on Pip himself—Miss Havisham's death has the quality of a displaced and, in the process, magnified act of retribution. Here, however, there is no Orlick to act as Pip's surrogate.[71] Miss Havisham's death renders more deeply problematic than any of those earlier scenes the relationship between action and agency, and in doing so brings to a head problems that have been steadily accumulating throughout Pip's narrative. As he leaves Satis House, Pip turns to look back at the place and is startled by the vivid return of his childhood delusion that Miss Havisham is 'hanging to the beam' of the old brewery (397). The strange choice of preposition (not 'from' the beam, but 'to' it) indicates an uncertain relationship to the image of retribution—a distressed confusion about whether this woman is a criminal or a victim—which is more than borne out in the scene that follows. Returning to the house, Pip finds Miss Havisham

seated in the ragged chair upon the hearth close to the fire, with her back towards me. In the moment when I was withdrawing my head to go quietly away, I saw her running at me shrieking, with a whirl of fire blazing all about her, and soaring at least as many feet above her head as she was high.

I had a double-caped great-coat on, and over my arm another thick coat. That I got them off, closed with her, threw her down, and got them over her; that I dragged the great cloth from the table for the same purpose, and with it dragged the heap of rottenness in the midst, and all the ugly things that were there; that we were on the ground struggling like desperate enemies, and that the closer I covered her, the more wildly she shrieked and tried to free herself; that this occurred I knew through the result, but not through anything that I felt, or thought, or knew that I did.   (397)

When Pip wrestles with the burning body on the floor, echoes of Compeyson's earlier struggle with Magwitch are unmistakable. But

[71] See particularly A. L. French, 'Beating and Cringing: *Great Expectations*', *Essays in Criticism*, 24 (1974), 147–68: 165–6.

it is unclear which role is Pip's—and unclear also how far this scene can bear the weight of meaning attached to it. Miss Havisham is too old a victim, and too sad, to sustain the weight of so much retribution, and Pip's account of her burning is unable to resolve its own blurred sense of where this violence is coming from and what is its object. Pip is Miss Havisham's enemy, throwing her to the ground, closing not 'on her' but 'with her' as Magwitch closed with Compeyson, holding her down 'like a prisoner who might escape'; but he is also her rescuer, protecting her; and her victim, burnt by contact with her. When they lie on the floor together, while the great patches of tinder still alight float down around them, the scene becomes a ghastly parody of a wedding consummation, caught between murderous rape and fulfilled love.

Certainly something remains unpurged about Miss Havisham even when the flames are put out. Was Miss Havisham Pip's persecutor, or was she, like him, a victim of other people's plotting? Does she offer a unifying symbol for the desolation of his expectations, or is she a misleading figure, a false authoress who has to be purged if he is to write the narrative of his life correctly? In the end, Miss Havisham seems most important for his narrative precisely because she will not allow those questions to be given simple answers. More urgently than any of the love-mad women in fiction before her, Miss Havisham raises the question of what it means to adopt another person's suffering as a means of representing and understanding one's own distressed relationship to the world. In what sense can one person's suffering ever be commensurate with another's? How do we weigh pain? Miss Havisham, like all the love-mad women before her, lives in the realm of hyperbole, as if only excess can do justice to her. Where Dickens's novel is, finally, subtler than its narrator is in allowing room for the recognition that, try as he does to subdue her narrative to the condition of a symbol—a metaphorical equivalent to the failure of his hopes—the violence necessary to effect that subordination is itself inordinate. Dying slowly amid the blackened ruins of her bridal feast, Miss Havisham preserves a damaged likeness of herself which, for all Pip's desire to repudiate her, commands his homage: 'Though every vestige of her dress was burnt, as they told me, she still had something of her old ghastly bridal appearance; for, they had covered her with white cotton-wool, and she lay with a white sheet loosely covering that . . .' (398).

For Collins and for Dickens, writing at a time when powerful sectors of the medical profession were increasingly insisting that knowledge of human life would be grounded in physiology rather than psychology, the love-mad woman provided a means of asking what constituted 'real' pain. Such women had always raised the question of 'truth'. Because they have suffered directly by men's falsehood or, indirectly, by the world's inability to guarantee their happiness, they prompt doubt about what *can* be assured in life. Couching their stories in hyperbole, they nevertheless resist the equation of excess with lying. Although they convey other stories within nineteenth-century fiction—and, in the process, come to mean both more and less than their own desolation—they still command pathos in themselves, even in the unsentimental world of the 1860s. Pip's parting view of Miss Havisham might stand as an epitaph for all her predecessors in fiction—a suitably ambiguous one which recognizes her power to return, even as it mourns her passing: shrouded in white, she has about her 'the phantom air of something that had been and was changed'.

Select Bibliography

Many of the books listed here are interdisciplinary in their approach. Rather than duplicate bibliographical entries, I have categorized the books as 'literary', 'historical' or 'theoretical' according to their primary use in this book. In several cases, this was necessarily to some degree arbitrary. Modern critical or facsimile editions of primary texts have been cited, where available. Elsewhere, the first edition has been used.

Novels and Other Primary Literary Sources

ANON., *Lady Cheveley; or, The Woman of 'Honour'. A New Version of Cheveley, the Man of Honour* (London, 1839).

AUSTEN, JANE, *Sense and Sensibility* (1811), ed. with intro. by Tony Tanner (Harmondsworth, 1969).

—— *Pride and Prejudice* (1813), ed. James Kingsley, new intro. by Isobel Armstrong (Oxford, 1990).

—— *Persuasion* (1817), ed. John Davie, new intro. by Claude Rawson (Oxford, 1990).

—— *Jane Austen: Selected Letters 1796–1817*, ed. R. W. Chapman (London, 1955; repr. Oxford, 1985).

—— *Volume the Second*, ed. B. C. Southam (Oxford, 1963).

—— *Volume the First*, ed. R. W. Chapman (London, 1984).

BLAIR, HUGH, *Lectures on Rhetoric and Belles Lettres*, 2 vols. (London, 1783).

BRADDON, MARY ELIZABETH, *Lady Audley's Secret*, 3 vols. (London, 1861).

BRONTË, CHARLOTTE ('Currer Bell'), *Jane Eyre* (1847), ed. Jane Jack and Margaret Smith (Oxford, 1969).

—— *Shirley* (1849), ed. Herbert Rosengarten and Margaret Smith (Oxford, 1979).

—— *Villette* (1853), ed. Herbert Rosengarten and Margaret Smith (Oxford, 1984).

—— *The Professor* (1857), ed. Margaret Smith and Herbert Rosengarten (Oxford, 1987).

—— *An Edition of the Early Writings of Charlotte Brontë*, 2 vols., ed. Christine Alexander (Oxford, 1987).

BRONTË, EMILY ('Ellis Bell'), *Wuthering Heights* (1847), ed. Hilda Marsden and Ian Jack (Oxford, 1976).

BROOKS, SHIRLEY, *The Gordian Knot: A Story of Good and Evil* (London, 1860).

BURY, LADY CHARLOTTE, *Self-Indulgence: A Tale of the Nineteenth Century* (Edinburgh, 1812).

BYRON, GEORGE GORDON, *Lord Byron: The Complete Poetical Works*, ed. Jerome J. McGann, 7 vols. (Oxford, 1980–93).

COCKTON, HENRY, *The Life and Adventures of Valentine Vox, the Ventriloquist* (London, 1844).

COLERIDGE, SAMUEL TAYLOR, *Biographia Literaria*, in Kathleen Coburn (gen. ed.), *The Collected Works of Samuel Taylor Coleridge* (Princeton, 1976–), vii, ed. James Engell and W. Jackson Bate.

COLLINS, WILKIE, *The Woman in White* (1860), ed. Harvey Peter Sucksmith (London, 1975).

——*Armadale* (1866), ed. with intro. by John Sutherland (Harmondsworth, 1995).

COWPER, WILLIAM, 'The Task', in *Cowper: Poetical Works*, ed. H. S. Milford, 4th edn., with corrections and additions by Norma Russell (London, 1967).

CUNNINGHAM, PETER, *The Story of Nell Gwyn; and, The Sayings of Charles II* (London, 1852).

DACRE, CHARLOTTE ('Rosa Matilda'), *Confessions of the Nun of St Omer: A Tale* (1805), fac. edn., 3 vols. in 2, with intro. by Devendra P. Varma (New York, 1972).

——*The Hours of Solitude: A Collection of Original Poems* (London, 1805).

——*Zofloya; or, The Moor* (1806), fac. edn., 4 vols., foreword by G. W. Knight, intro. by Devendra P. Varma (New York, 1974).

——*The Libertine* (1807), fac. edn., 4 vols., foreword by Sandra Knight-Roth, intro. by Devendra P. Varma (New York, 1974).

——*The Passions* (1811), fac. edn., 4 vols., foreword by Sandra Knight-Roth, intro. by Devendra P. Varma (New York, 1974).

DAVENANT, WILLIAM, *The Dramatic Works of Sir William D'Avenant*, 5 vols., ed. James Maidment and W. H. Logan (Edinburgh, 1872–4).

DICKENS, CHARLES, *Bleak House* (1853), ed. Norman Page, intro. by J. Hillis Miller (Harmondsworth, 1971).

——'Where We Stopped Growing', *Household Words*, 1 Jan. 1853, 361–3.

——*Hard Times* (1854), ed. Paul Schlicke (Oxford, 1989).

——*Great Expectations* (1861), ed. Margaret Cardwell (Oxford, 1993).

——*The Letters of Charles Dickens*, 3 vols., ed. Walter Dexter (London, 1938).

——*Speeches of Charles Dickens*, ed. K. J. Fielding (Oxford, 1960).

——(attrib.), 'The Dirty Old Man: A Lay of Leadenhall', *Household Words*, 8 Jan. 1853, 396–7.

[——and COLLINS, WILKIE], 'The Lazy Tour of Two Idle Apprentices', *Household Words*, Oct. 1857, 313–19, 334—49, 361–7, 385–93, 409–16.

——and WILLS, W. H., 'A Curious Dance Round a Curious Tree', *Household Words*, 17 Jan. 1852, 385–9.

EVANS, THOMAS (ed.), *Old Ballads, Historical and Narrative, with Some of Modern Date* (London, 1784).

GASKELL, MRS [ELIZABETH], *Life of Charlotte Brontë* (London, 1857).

——*Cousin Phyllis and Other Tales* (London, 1865).

GILBERT, WILLIAM, *Shirley Hall Asylum; or, The Memoirs of a Monomaniac* (London, 1863).

GODWIN, WILLIAM, *Fleetwood; or, The New Man of Feeling*, 3 vols. (London, 1805).

HAMILTON, ELIZABETH, *Memoirs of Modern Philosophers*, 3 vols. (Bath, 1800).

HAYS, MARY, *Memoirs of Emma Courtenay*, 2 vols. (London, 1796).

HELME, ELIZABETH, *The Farmer of Inglewood Forest: A Novel*, 4 vols. (London, 1796).

HOGG, JAMES, *The Private Memoirs and Confessions of a Justified Sinner* (Edinburgh, 1824).

——'Strange Letter of a Lunatic', *Fraser's Magazine*, 2 (1830), 526–32.

HOLCROFT, THOMAS, *Anna St Ives* (London, 1792).

——*The Adventures of Hugh Trevor* (London, 1794–7).

JERROLD, DOUGLAS, *Nell Gwynne; or, The Prologue, A Comedy in Two Acts* (London, 1833).

JEWSBURY, GERALDINE ENDSOR, *Constance Herbert*, 3 vols. (London, 1855).

KERRIGAN, JOHN (ed.), *Motives of Woe: Shakespeare and 'Female Complaint'. A Critical Anthology* (Oxford, 1991).

LAMB, LADY CAROLINE, *Glenarvon*, 3 vols. (London, 1816).

——*Glenarvon*, rev. edn., 3 vols. (London, 1816).

LEFANU, J. SHERIDAN, *Uncle Silas*, 3 vols. (London, 1864).

——*The Rose and the Key*, 3 vols. (London, 1871).

LEWIS, MATTHEW, *Poems* (London, 1812).

LILLO, GEORGE, *The London Merchant; or, The History of George Barnwell: A Tragedy* (London, 1731).

LINDSAY, JACK (ed.), *Loving Mad Tom: Bedlamite Verses of the XVI and XVII Centuries*, illustrations by Norman Lindsay, foreword by Robert Graces, musical transcriptions by Peter Warlock (London, 1927).

Lytton, Edward Bulwer-, *Godolphin*, 3 vols. (London, 1833).

——*Lucretia; or, The Children of Night*, 3 vols. (London, 1846).

——*The Caxtons: A Family Romance*, 3 vols. (London, 1849).

——*Not So Bad as We Seem; or, Many Sides to a Character* (London, 1851).

Lytton, Rosina ('Lady Lytton Bulwer'), *Cheveley; or, The Man of Honour*, 3 vols. (London, 1839).

——*A Blighted Life: A True Story* (1880), fac. edn. with intro. by Marie Mulvey Roberts (Bristol, 1994).

McGann, Jerome J. (ed.), *Oxford Anthology of Romantic Period Verse* (Oxford, 1993).

Mackenzie, Henry, *The Man of Feeling* (1771), ed. with intro. by Brian Vickers (Oxford, 1967).

Maturin, Charles, *The Wild Irish Boy*, 3 vols. (London, 1808).

——*The Milesian Chief: A Romance*, 4 vols. (London, 1812).

——*Melmoth: The Wanderer* (1820), ed. with intro. by Douglas Grant (London, 1968).

——*Connal; ou, Les Milesians*, traduit de l'anglais par Mme La Comtesse [Mole] (Paris, 1828).

[Maxwell, W. H.], *O'Hara; or, 1798*, 2 vols. (London, 1825).

Meredith, George, *The Ordeal of Richard Feverel: A History of Father and Son*, 3 vols. (London, 1859).

Morgan, Lady Sydney, *St Clair; or, The Heiress of Desmond* (London, 1803).

Payn, James, *Lost Sir Massingberd*, 3 vols. (London, 1864).

Pickering, Ellen, *Nan Darrell; or, The Gipsy Mother*, 3 vols. (London, 1839).

Prest, Thomas Peckett, *The Maniac Father; or, The Victim of Seduction. A Romance* (London, 1842).

Puttenham, George, *The Arte of English Poesie* (1594), ed. Gladys Doidge Willcock and Alice Walker (Cambridge, 1936).

Reade, Charles, *Hard Cash*, 3 vols. (London, 1863).

——*A Terrible Temptation*, 3 vols. (London, 1871).

——and Boucicault, Dion, *Foul Play*, 3 vols. (London, 1868).

Rhys, Jean, *Wide Sargasso Sea* (London, 1966).

Ruskin, John, *The Stones of Venice*, 3 vols. (London, 1851–6).

Scott, Sir Walter, *Guy Mannering; or, The Astrologer*, 3 vols. (Edinburgh, 1815).

——*The Bride of Lammermoor* (1819; rev. and expanded 1830; Oxford, 1991).

——*The Visionary* (1819), ed. with intro. by Peter Garside (Cardiff, 1984).

——*Ivanhoe* (1820), ed. A. N. Wilson (Harmondsworth, 1982).

——*St Ronan's Well*, 3 vols. (Edinburgh, 1823).

——*The Letters of Sir Walter Scott*, ed. H. J. C. Grierson *et al.*, 12 vols. (London, 1932–7).

SHAKESPEARE, WILLIAM, *The Riverside Shakespeare*, ed. G. Blakemore Evans *et al.* (Boston, 1974).

SHELLEY, PERCY BYSSHE, *The Complete Works of Percy Bysshe Shelley*, ed. Neville Rogers, 4 vols. (Oxford, 1975).

STANHOPE, LOUISA SIDNEY, *Madelina: A Tale, Founded on Facts*, 4 vols. (London, 1814).

STERNE, LAURENCE, *The Life and Opinions of Tristram Shandy* (1760–1), ed. Graham Petrie, intro. by Christopher Ricks (Harmondsworth, 1967).

——*A Sentimental Journey through France and Italy by Mr Yorick* (1768), ed. Ian Jack (London, 1968).

SURTEES, ROBERT S., *Handley Cross; or, The Spa Hunt*, 3 vols. (London, 1843).

THACKERAY, WILLIAM MAKEPEACE, *Vanity Fair* (1848), ed. with intro. by John Sutherland (Oxford, 1983).

——*The Oxford Thackeray*, viii: *Miscellaneous Contributions to Punch 1843–1854*, ed. George Saintsbury (London, 1908).

——*The Letters and Private Papers of William Makepeace Thackeray*, 4 vols., ed. Gordon N. Ray (London, 1945–6).

TROLLOPE, FRANCES, *The Life and Adventures of Michael Armstrong, The Factory Boy*, 3 vols. (London, 1839).

WARREN, SAMUEL, *Ten Thousand a Year*, 3 vols. (London, 1841).

WOLLSTONECRAFT, MARY, *The Wrongs of Woman; or, Maria. A Fragment* (1798), in *The Works of Mary Wollstonecraft*, ed. Janet Todd and Marilyn Butler, 7 vols. (London, 1989).

WOOLF, VIRGINIA, *Between the Acts* (London, 1941).

YEATS, W. B., *W. B. Yeats: The Poems*, ed. Richard J. Finneran, corr. repr. (London, 1984).

Primary Medical Sources and Related Material

ANON., 'M.D. and MAD', *All the Year Round*, 22 Feb. 1862, 510–13.

ARNOLD, THOMAS, *Observations on the Nature, Kinds, Causes, and Prevention of Insanity, Lunacy, or Madness*, 2 vols. (Leicester, 1782–6).

BAKEWELL, THOMAS, *The Domestic Guide in Cases of Insanity: Pointing Out the Causes, Means of Preventing, and Proper Treatment of that Disorder* (Hanley, 1805).

BATTIE, WILLIAM, *A Treatise on Madness* (London, 1758).

BLACK, W., *A Dissertation on Insanity, Illustrated with Tables, and Extracted from Between Two and Three Thousand Cases in Bedlam* (London, 1810).

BROWNE, W. A. F., *What Asylums Were, Are and Ought to Be: Being the Substance of Five Lectures Delivered Before the Managers of the Montrose Royal Lunatic Asylum* (Edinburgh, 1837).

BUCKNILL, JOHN CHARLES, and TUKE, DANIEL H., *A Manual of Psychological Medicine: Containing the History, Nosology, Description, Statistics, Diagnosis, Pathology, and Treatment of Insanity. With an Appendix of Cases* (London, 1858).

BURROWS, GEORGE MANN, *An Inquiry into Certain Errors Relative to Insanity; and their Consequences; Physical, Moral and Civil* (London, 1820).

——*Commentaries on the Causes, Forms, Symptoms, and Treatment, Moral and Medical, of Insanity* (London, 1828).

——*A Letter to Sir Henry Halford, Bart, K.C.H., &c., Touching Some Points of the Evidence, and Observations of Counsel, on a Commission of Lunacy on Mr. Edward Davies* (London, 1830).

BURTON, ROBERT, *The Anatomy of Melancholy* (1621), 3 vols., ed. Thomas C. Faulkner, Nicolas K. Kiessling, and Rhonda Blair (Oxford, 1989–94).

CARTER, ROBERT BRUDENELL, *On the Pathology and Treatment of Hysteria* (London, 1853).

CHEYNE, GEORGE, *The English Malady; or, A Treatise of Nervous Diseases of all Kinds, as Spleen, Vapours, Lowness of Spirits, Hypochondriacal, and Hysterical Distempers, &c.* (London, 1733).

CHURCHILL, FLEETWOOD, *Outlines of the Principal Diseases of Females* (Dublin, 1838).

CONOLLY, JOHN, *An Inquiry Concerning the Indications of Insanity with Suggestions for the Better Protection and Care of the Insane* (1830), repr. with intro. by Richard Hunter and Ida Macalpine (London, 1964).

——*On the Construction and Government of Lunatic Asylums and Hospitals for the Insane* (1847), repr. with intro. by Richard Hunter and Ida Macalpine (London, 1968).

—— *The Treatment of the Insane without Mechanical Restraints* (1856), fac. edn. with intro. by Richard Hunter and Ida Macalpine (Folkstone, 1973).

COX, JOSEPH MASON, *Practical Observations on Insanity; In which some Suggestions are Offered towards an Improved Mode of Treating Diseases of the Mind, and some Rules Proposed which it is Hoped May Lead to a More Humane and Successful Method of Cure: To which are Subjoined, Remarks on Medical Jurisprudence as Connected with Diseased Intellect* (London, 1804)

—— *Practical Observations on Insanity*, 2nd edn., corr. and greatly enlarged (London, 1806).

—— *Practical Observations on Insanity as it Relates to Diseased Intellect*, 3rd edn., corr. and greatly enlarged (London, 1813).

CRICHTON, SIR ALEXANDER, *An Inquiry into the Nature and Origin of Mental Derangement*, 2 vols. (London, 1798).

CROWTHER, BRYAN, *Practical Remarks on Insanity: To which is Added a Commentary on the Dissection of the Brains of Maniacs* (London, 1811).

CULLEN, WILLIAM, *First Lines of the Practice of Physic*, 4 vols. (Edinburgh, 1776–84).

DARWIN, CHARLES, *On the Origin of Species by Means of Natural Selection; or, The Preservation of Favoured Races in the Struggle for Life* (London, 1859).

ELLIS, S. C., *A Treatise on the Nature, Symptoms, Causes and Treatment of Insanity* (London, 1838).

ESQUIROL, JEAN-ÉTIENNE DOMINIQUE, 'Érotomanie', in *Dictionaire des sciences médicales, par une société de médecins et de chirurgiens*, 60 vols. (Paris, 1812–22), xiii.

——*Des maladies mentales* (Paris, 1838).

——*Mental Maladies: A Treatise on Insanity*; trans. with additions by E. K. Hunt (London, 1845).

FERRIAR, JOHN, *Medical Histories and Reflections*, 3 vols. (London, 1795–8).

FORBES, J., TWEEDIE, A., and CONOLLY, J., *The Cyclopaedia of Practical Medicine*, 4 vols. (London, 1833–5).

GRAY, JONATHAN, *A History of the York Lunatic Asylum* (York, 1815).

HALLARAN, WILLIAM SAUNDERS, *An Enquiry into the Causes Producing the Extraordinary Addition to the Number of Insane, together with Extended Observations on the Cure of Insanity; with Hints as to the Better Management of Public Asylums for Insane Persons . . .* (Cork, 1810).

——*Practical Observations on the Causes and Cure of Insanity* (Cork, 1818).

HALLIDAY, ANDREW, *A General View of the Present State of Lunatics and Lunatic Asylums in Great Britain and Ireland* (London, 1828).

HARPER, ANDREW, *A Treatise on the Real Cause and Cure of Insanity, in which the Nature and Distinctions of this Disease are Fully Explained, and the Treatment Established on New Principles* (London, 1789).

HASLAM, JOHN, *Observations on Madness and Melancholy; Including Practical Remarks on those Diseases; together with Cases; and an Account of the Morbid Appearances on Dissection*, 2nd edn., enlarged (London, 1809).

——*Illustrations of Madness: Exhibiting a Singular Case of Insanity, and a No Less Remarkable Difference in Medical Opinion Developing the Nature of Assailment, and the Manner of Working Events; with a Description of the Tortures Experienced by Bomb-Bursting, Lobster-Cracking, and Lengthening the Brain Embellished with a Curious Plate* (London, 1810).

——*Considerations on the Moral Management of Insane People* (London, 1817).

——*Medical Jurisprudence as it Relates to Insanity According to the Laws of England* (London, 1817).

HAWKES, JOHN, 'On the Increase of Insanity', *Journal of Psychological Medicine and Mental Pathology*, 10 (1857), 508–21.

JAMES, ROBERT, *A Medicinal Dictionary*, 3 vols. (London, 1743–5).

JONES, ROBERT, *An Enquiry into the Nature, Causes and Termination of Nervous Fevers; Together with Observations Tending to Illustrate the Method of Restoring His Majesty to Health and of Preventing Relapses of his Disease* (London, 1789).

LOCKE, JOHN, *An Essay Concerning Human Understanding* (London, 1690).

LYTTON, ROBERT, Letter to the *Daily Telegraph*, 19 July 1858, 5.

MASON, SAMUEL, *The Philosophy of Female Health: Being an Enquiry into its Connections with, and Dependence upon, the Due Performance of Uterine Functions; with Observations on the Nature, Causes, and Treatment of Female Disorders in General* (London, 1845).

MAUDSLEY, HENRY, *The Physiology and Pathology of the Mind* (London, 1867).

——*Sex in Mind and Education* (New York, 1874).

——*Body and Will: In its Metaphysical, Psychological and Pathological Aspects* (London, 1883).

——*The Pathology of Mind: A Study of its Distempers, Deformities and Disorders* (1895), repr. with intro. by Aubrey Lewis (London, 1979).

MEAD, RICHARD, *Monita et praecepta medica* (1751); *Medical Precepts and Cautions*, trans. Thomas Stack (London, 1751).

MIDRIFF, J., *Observations on the Spleen and Vapours, Containing Remarkable Cases of Persons of Both Sexes and All Ranks from the Aspiring Directors to the Humbler Bubbler who have been Miserably Afflicted with these Melancholy Disorders since the Fall of the South Sea and Other Public Stocks* (London, 1721).

MITFORD, JOHN ('Alfred Burton'), *A Description of the Crimes and Horrors in the Interior of Warburton's Private Mad-House at Hoxton, Commonly Called Whitmore House* (London, 1822).

——*Part Second of the Crimes and Horrors of the Interior of Warburton's Private Mad-Houses at Hoxton and Bethnal Green and of These Establishments in General with Reasons for Their Total Abolition* (London, 1825).

MORISON, ALEXANDER, *Outlines of Lectures on Mental Diseases* (Edinburgh, 1825).

——*Outlines of Lectures on Mental Diseases*, 3rd edn. (London, 1826).

——*The Physiognomy of Mental Diseases* (London, 1840).

——*Outlines of Lectures on the Nature, Causes and Treatment of Insanity*, 4th edn., ed. Thomas Coutts Morison (London, 1848).

——*Lectures on Insanity*, 5th edn., ed. Thomas Coutts Morison (Edinburgh, 1856).

PERCEVAL, JOHN, *A Narrative of the Treatment Experienced by a Gentleman, during a State of Mental Derangement, Designed to Explain the Causes and Nature of Insanity, and to Expose the Injudicious Conduct toward Many Unfortunate Sufferers under that Calamity*, 2 vols. (London, 1838–40).

PINEL, PHILIPPE, *Traité médico-philosophique sur l'aliénation mentale; ou, La Manie* (Paris, 1801).

——*A Treatise on Insanity*, trans. D. D. Davis (Sheffield, 1806).

——*Traité médico-philosophique sur l'aliénation mentale*, 2ᵉ édition, entièrement refondue et très augmentée (Paris, 1809).

PRICHARD, J. C., *A Treatise on Insanity*, offprinted from J. Forbes, A. Tweedie, and J. Conolly (eds.), *The Cyclopaedia of Practical Medicine* (London, 1833).

——*A Treatise on Insanity* (London, 1835).

READ, JOHN, *Essays on Insanity, Hypochondria, and other Nervous Affections* (London, 1816).

RUSH, BENJAMIN, *An Inquiry into the Influence of Physical Causes upon the Moral Faculty* (Philadelphia, 1786).

——*Medical Inquiries and Observations Upon the Diseases of the Mind*, 2nd edn. (Philadelphia, 1805).

TEMPEST, JOHN, *Narrative of the Treatment Experienced by John Tempest, Esq., of Lincoln's Inn, Barrister of Law, During Fourteen Months Solitary Confinement under a False Imputation of Lunacy* (London, 1830).

TROTTER, THOMAS, *A View of the Nervous Temperament: Being a Practical Enquiry into the Increasing Prevalence of those Diseases Commonly Called Nervous* (London, 1807).

TUKE, SAMUEL, *Description of the Retreat, an Institution near York, for Insane Persons of the Society of Friends* (York, 1813).

TYNDALL, JOHN, *Essays on the Use and Limit of the Imagination in Science* (London, 1870).

——*Hours of Exercise in the Alps* (London, 1871).

——*Address Delivered Before the British Association Assembled at Belfast* (London, 1874).

Literary Criticism, Biography, and Publishing History

ALEXANDER, CHRISTINE, *The Early Writings of Charlotte Brontë* (Oxford, 1983).

ALLOTT, MIRIAM (ed.), *The Brontës: The Critical Heritage* (London, 1974).

ALTICK, RICHARD D., *The English Common Reader: A Social History of the Mass Reading Public 1800–1900* (Chicago, 1957).

—— *Paintings from Books: Art and Literature in Britain 1760–1900* (Columbus, Oh., 1985).

ANDERSON, JAMES, *Sir Walter Scott and History, with Other Papers* (Edinburgh, 1981).

ANON., Review of Charlotte Dacre, *The Passions*, *The Critical Review*, 3rd ser., 24 (1811), 550.

—— *Catalogue of the Public Library, Broad-Street, Aberdeen* (Aberdeen, 1821).

—— Review of Edward Bulwer-Lytton, *Lucretia; or, The Children of Night*, *The Athenaeum*, 5 Dec. 1846, 1240–2.

—— 'The Province of Tragedy: Bulwer and Dickens', *Westminster and Foreign Quarterly Review*, 47 (Apr. 1847), 1–11.

ARMSTRONG, NANCY, *Desire and Domestic Fiction: A Political History of the Novel* (New York, 1987).

AUERBACH, NINA, *Woman and the Demon: The Life of a Victorian Myth* (Cambridge, Mass., 1982).

BABB, LAWRENCE, *The Elizabethan Malady: A Study of Melancholia in English Literature from 1580 to 1642* (East Lansing, Mich., 1951).

BAINE, R. M., *Thomas Holcroft and the Revolutionary Novel* (Athens, Ga., 1965).

BARNES, JOHN, *Authors, Publishers and Politicians: The Quest for an Anglo-American Copyright Agreement 1815–1854* (London, 1974).

—— 'Depression and Innovation in the British and American Booktrade 1819–1939', in Robin Myers and Michael Harris (eds.), *Economics of the British Booktrade 1605–1939* (Cambridge, 1985).

BEER, GILLIAN, *Darwin's Plots: Evolutionary Narrative in Darwin, George Eliot and Nineteenth-Century Fiction* (London, 1983).

—— ' "Our Unnatural No-Voice": The Heroic Epistle, Pope, and Women's Gothic', *Yearbook of English Studies*, Special Number, 12, 'Heroes and the Heroic', ed. G. K. Hunter and C. J. Rawson (1982), 125–51.

—— 'Beyond the Sensible, Lucid Jane', *Times Higher Education Supplement*, 16 Jan. 1987, 17.

BESTERMAN, THEODORE (ed.), *The Publishing Firm of Cadell & Davies: Select Correspondence and Accounts 1793–1836* (Oxford, 1938).

BJÖRK, HARRIET, *The Language of Truth: Charlotte Brontë, the Woman Question and the Novel* (Lund, 1974).

BOLTON, H. PHILIP, *Scott Dramatized* (London, 1992).

BOSTETTER, E. E., 'The Original Della Cruscans and the Florence Miscellany', *Huntington Library Quarterly*, 19 (1956), 277–300.

BOUMELHA, PENNY, *Charlotte Brontë* (Hemel Hempstead, 1990).

BRADBROOK, FRANK W., *Jane Austen and her Predecessors* (Cambridge, 1967).

BRIGHTFIELD, MIRON, *Victorian England and its Novels 1840–1870*, 4 vols. (Los Angeles, 1968).

BROOKS, PETER, *Reading for the Plot: Design and Intention in Narrative* (Oxford, 1984).

BROWN, DAVID, *Sir Walter Scott and the Historical Imagination* (London, 1979).

BUTLER, MARILYN, *Jane Austen and the War of Ideas* (Oxford, 1975; rev. edn. 1987).

—— *Romantics, Rebels and Reactionaries: English Literature and its Background 1760–1830* (Oxford, 1981).

CAMERON, KENNETH NEIL, *The Young Shelley: Genesis of a Radical* (New York, 1950).

CAMPBELL, JAMES L., SR., *Edward Bulwer-Lytton* (Boston, 1986).

CHASE, KAREN, *Eros and Psyche: The Representation of Personality in Charlotte Brontë, Charles Dickens, George Eliot* (London, 1984).

CHRISTENSEN, ALLAN, *Edward Bulwer-Lytton: The Fiction of New Regions* (Athens, Ga., 1976).

CLARK, TIMOTHY, *Embodying Revolution: The Figure of the Poet in Shelley* (Oxford, 1989).

COOPER, ANDREW M., 'Blake and Madness: The World Turned Inside Out', *English Literary History*, 57 (1990), 585–642.

COTTOM, DANIEL, *The Civilised Imagination: A Study of Ann Radcliffe, Jane Austen, and Sir Walter Scott* (London, 1985).

DAICHES, DAVID, *Literary Essays* (Edinburgh, 1956).

DAVIES, NEIL PHARR, *The Life of Wilkie Collins* (Urbana, Ill., 1956).

DEVEY, LOUISA, *Life of Rosina, Lady Lytton, with Numerous Extracts from her Autobiography and Other Original Documents. Published in Vindication of her Memory* (London, 1887).

DIJKSTRA, BRAM, *Idols of Perversity: Fantasies of Feminine Evil in Fin-de-Siècle Culture* (Oxford, 1986).

DOOLEY, ALLAN C., *Author and Printer in Victorian England* (Charlottesville, Va., 1992).

DREW, PHILIP, 'Past Imagining', *Times Literary Supplement*, 20–6 Apr. 1990, 422.

DUFFY, EDWARD, *Rousseau in England: The Context for Shelley's Critique of the Enlightenment* (Berkeley, 1979).

EAGLETON, TERRY, *Myths of Power: A Marxist Study of the Brontës* (London, 1975).

EIGNER, EDWIN M., 'Bulwer's Accommodation to the Realists', in Harold
Orel and George W. Worth (eds.), *The Nineteenth-Century Writer and
his Audience* (Lawrence, Kan., 1969).

ELWIN, MALCOLM, *Charles Reade: A Biography* (London, 1931).

ESCOTT, T. H. S., *Edward Bulwer First Baron Lytton of Knebworth: A
Social, Personal, and Political Monograph* (London, 1910).

FARRELL, JOHN P., *Revolution as Tragedy: The Dilemma of the Moderate
from Scott to Arnold* (Ithaca, NY, 1980).

FEATHER, JOHN, *A History of British Publishing* (London, 1988).

FEDER, LILLIAN, *Madness in Literature* (Princeton, 1980).

FERGUS, JAN, *Jane Austen and the Didactic Novel* (Totowa, NJ, 1983).

FERRER, DANIEL, *Virginia Woolf and the Madness of Language*, trans.
Geoffrey Bennington and Rachel Bowlby (London, 1990).

FRANKLIN, CAROLINE, 'Feud and Faction in *The Bride of Lammermoor*',
Scottish Literary Journal, 14 (1987), 18–31.

GALLAGHER, CATHERINE, *The Industrial Reformation of English
Fiction: Social Discourse and Narrative Form 1832–1867* (Chicago,
1985).

GALLAWAY, W. F., 'The Conservative Attitude toward Fiction 1770–1830',
PMLA, 55 (1940), 1041–59.

GARSIDE, PETER DIGNUS, 'Union and *The Bride of Lammermoor*', *Studies
in Scottish Literature*, 19 (1984), 72–93.

—— 'J. F. Hughes and the Publication of Popular Fiction 1803–1810', *The
Library*, 6th ser., 9 (1987), 240–58.

GILBERT, SANDRA M., and GUBAR, SUSAN, *The Madwoman in the Attic:
The Woman Writer and the Nineteenth-Century Literary Imagination*
(New Haven, 1979).

GLEN, HEATHER, Introduction to Charlotte Brontë, *The Professor*
(London, 1989).

GOLDBERG, RITA, *Sex and Enlightenment: Women in Richardson and
Diderot* (Cambridge, 1984).

GORDON, CATHERINE M., *British Paintings of Subjects from the English
Novel 1740–1870* (New York, 1988).

GORDON, ROBERT C., '*The Bride of Lammermoor*: A Novel of Tory
Pessimism', *Nineteenth Century Fiction*, 12 (1957), 110–24.

—— and HOOK, ANDREW, Exchange of letters, *Nineteenth Century
Fiction*, 13 (1969), 493–9.

GRAHAM, PETER W., *Don Juan and Regency England* (Charlottesville, Va.,
1990).

GRAHAM, WALTER, *The Beginnings of the English Literary Periodical: A
Study of Periodical Literature 1665–1715* (New York, 1926).

GREGORY, ALLENE, *The French Revolution and the English Novel* (New
York, 1915).

GROB, SHIRLEY, 'Dickens and Some Motifs of the Fairy Tale', *Texas Studies
in Literature and Language*, 5 (1965), 567–79.

GROSS, JOHN, and PEARSON, GABRIEL (eds.), *Dickens and the Twentieth Century* (London, 1962).

GRUDIN, PETER, 'Jane and the Other Mrs Rochester: Excess and Restraint in *Jane Eyre*', *Novel*, 10 (1977), 145–57.

HAGSTRAM, JEAN H., *Sex and Sensibility: Ideal and Erotic Love from Milton to Mozart* (Chicago, 1980).

HARGREAVES-MAWDSLEY, W. N., *The English Della Cruscans* (The Hague, 1967).

HARRIS, JOHN B., *Charles Robert Maturin: The Forgotten Imitator* (New York, 1980).

HONAN, PARK, *Jane Austen: Her Life* (London, 1987).

HOOK, ANDREW, '*The Bride of Lammermoor*: A Re-examination', *Nineteenth Century Fiction*, 22 (1967), 111–26.

HUGHES, WINIFRED, *The Maniac in the Cellar: Sensation Novels of the 1860s* (Princeton, 1980).

INGRAM, ALLAN, *The Madhouse of Language: Writing and Reading Madness in the Eighteenth Century* (London, 1991).

JACK, IAN, 'Physiognomy, Phrenology and Characterization in the Novels of Charlotte Brontë', *Brontë Society Transactions*, 15 (1970), 377–91.

JACKSON, T. A., *Charles Dickens: The Progress of a Radical* (London, 1837).

JERDAN, WILLIAM, Introductory article to the *Literary Gazette*, 2 Jan. 1819, 1.

JOHNSON, CLAUDIA L., 'A "Sweet Face as White as Death": Jane Austen and the Politics of Female Sensibility', *Novel*, 22 (1989), 159–74.

JOHNSON, JUDY VAN SICKLE, 'The Bodily Frame: Learning Romance in *Persuasion*', *Nineteenth Century Fiction*, 38 (1983), 43–61.

JOHNSTON, RUTH D., '*The Professor*: Charlotte Brontë's Hysterical Text, or Realistic Narrative and the Ideology of the Subject from a Feminist Perspective', *Dickens Studies Annual*, 18 (1989), 353–80.

JONES, ANN H., *Ideas and Innovations: Best Sellers of Jane Austen's Age* (New York, 1986).

KAPLAN, FRED, *Dickens and Mesmerism: The Hidden Springs of Fiction* (Princeton, 1975).

KERR, JAMES, *Fiction against History: Scott as Storyteller* (Cambridge, 1989).

KIRKHAM, MARGARET, *Jane Austen: Feminism and Fiction* (Totowa, NJ, 1983).

KNIGHT-ROTH, SANDRA, 'Charlotte Dacre and the Gothic Tradition', doctoral dissertation, Dalhousie University, Nova Scotia, 1972.

KRAMER, DALE, *Charles Robert Maturin* (New York, 1973).

KUCICH, JOHN, *Repression in Victorian Fiction: Charlotte Brontë, George Eliot, and Charles Dickens* (Berkeley, 1989).

LAUBER, JOHN, *Sir Walter Scott*, rev. edn. (Boston, 1989).

LEAVY, BARBARA FASS, 'Wilkie Collins's Cinderella: The History of Psychology and *The Woman in White*', *Dickens Studies Annual*, 10 (1982), 91–141.

LEVINE, GEORGE, 'Determinism and Responsibility in the Works of George Eliot', *PMLA*, 77 (1962), 268–79.

—— *The Realistic Imagination: English Fiction from Frankenstein to Lady Chatterley* (Chicago, 1981).

—— *Darwin and the Novelists: Patterns of Science in Victorian Fiction* (Boston, 1988).

—— (ed.), *One Culture: Essays in Science and Literature* (Madison, Wis., 1987).

LINDER, CYNTHIA A., *Romantic Imagery in the Novels of Charlotte Brontë* (London, 1978).

LIPKING, LAWRENCE, *Abandoned Women and the Poetic Tradition* (Chicago, 1988).

LOCKHART, JOHN GIBSON, *Memoirs of the Life of Sir Walter Scott, Bart.*, 2nd edn., 10 vols. (Edinburgh, 1839).

LOESBERG, JONATHAN, 'The Ideology of Narrative Form in Sensation Fiction', *Representations*, 13 (1986), 115–38.

LOWES, JOHN LIVINGSTONE, 'The Loveres Maladye of Hereos', *Modern Philology*, 11 (1913–14), 491–546.

LUKÁCS, GEORGE, *The Historical Novel*, trans. Hannah and Stanley Mitchell (London, 1962).

LYTTON, VICTOR ALEXANDER, 2ND EARL OF, *The Life of Edward Bulwer, First Lord Lytton, by his Grandson the Earl of Lytton* (London, 1913).

McCOMBIE, FRANK, 'Scott, *Hamlet*, and *The Bride of Lammermoor*', *Essays in Criticism*, 25 (1975), 419–36.

McKEON, MALCOLM, *The Origins of the English Novel 1600–1740* (Baltimore, 1987).

MacQUEEN, JOHN, *The Rise of the Historical Novel: The Enlightenment and Scottish Literature* (Edinburgh, 1982).

MARTIN, PHILIP, *Mad Women in Romantic Writing* (Brighton, 1987).

MAYNARD, JOHN, *Charlotte Brontë and Sexuality* (London, 1984).

MAYO, ROBERT, 'The Contemporaneity of the Lyrical Ballads', *PMLA*, 69 (1954), 486–522.

MEISEL, MARTIN, 'Miss Havisham Brought to Book', *PMLA*, 81/3 (1966), 278–85.

MELLOR, ANNE K. (ed.), *Romanticism and Feminism* (Bloomington, Ind., 1988).

MEYER, SUSAN, 'Colonialism and the Figurative Strategy of *Jane Eyre*', *Victorian Studies*, 33 (1990), 247–68.

MILES, ROBERT, *Gothic Writings 1750–1820: A Genealogy* (London, 1993).

MILLER, D. A., '*Cage aux folles*: Sensation and Gender in Wilkie Collins's

The Woman in White', *Representations*, 14 (1986), 107–36; repr. in D. A. Miller, *The Novel and the Police* (Berkeley, 1988).

MILLER, J. HILLIS, *Fiction and Repetition: Seven English Novels* (Oxford, 1982).

MILLER, KARL, *Doubles: Studies in Literary History* (Oxford, 1985).

MITCHELL, JEROME, *The Walter Scott Operas: An Analysis of Operas Based on the Works of Sir Walter Scott* (University, Alabama, 1977).

MOGLEN, HELENE, *Charlotte Brontë: The Self Conceived* (New York, 1976).

MONAGHAN, DAVID (ed.), *Jane Austen in a Social Context* (London, 1981).

MOYNAHAN, JULIAN, 'The Hero's Guilt: The Case of *Great Expectations*', *Essays in Criticism*, 10 (1960), 60–79.

MULLAN, JOHN, *Sentiment and Sociability: The Language of Feeling in the Eighteenth Century* (Oxford, 1988).

MUSSELWHITE, DAVID E., *Partings Welded Together: Politics and Desire in the Nineteenth-Century English Novel* (London, 1987).

MYER, VALERIE GROSVENOR, '*Jane Eyre*: The Madwoman as Hyena', *Notes and Queries*, NS 35 (1988), 318.

NESTOR, PAULINE, *Female Friendships and Communities: Charlotte Brontë, George Eliot, Elizabeth Gaskell* (Oxford, 1985).

O'HAYDEN, JOHN (ed.), *Scott: The Critical Heritage* (London, 1970).

OLSHIN, TOBY A., 'Jane Austen: A Romantic, Systematic, or Realistic Approach to Medicine?', *Studies in Eighteenth Century Culture*, 10 (1981), 313–26.

PAGE, NORMAN (ed.), *Wilkie Collins: The Critical Heritage* (London, 1974).

PARKER, PATRICIA, *Literary Fat Ladies: Rhetoric, Gender, Property* (London, 1987).

PARKER, WILLIAM, Preface to Sir Walter Scott, *The Bride of Lammermoor* (London, 1988).

PARSONS, COLEMAN O., 'The Dalrymple Legend in *The Bride of Lammermoor*', *Review of English Studies*, 73 (1943), 51–8.

PLANT, MARJORIE, *The English Book Trade: An Economic History of the Making and Sale of Books* (London, 1939).

PLUMMER, JOHN, 'The Original Miss Havisham', *The Dickensian*, 22 (1906), 298.

POLITI, JINA, *The Novel and its Presuppositions: Changes in the Conceptual Structure of Novels in the Eighteenth and Nineteenth Centuries* (Amsterdam, 1976).

POOVEY, MARY, *The Proper Lady and the Woman Writer: Ideology as Style in the Works of Mary Wollstonecraft, Mary Shelley and Jane Austen* (Chicago, 1984).

RAY, GORDON N., *Thackeray: The Uses of Adversity 1811–1846* (London, 1955).

RIGNEY, BARBARA HILL, *Madness and Sexual Politics in the Feminist Novel: Studies in Brontë, Woolf, Lessing and Atwood* (Madison, Wis., 1978).

RIMMON-KENAN, SHLOMITH, *Discourse in Psychoanalysis and Literature* (London, 1987).

ROBERTS, WARREN, *Jane Austen and the French Revolution* (London, 1979).

ROBERTSON, FIONA, Introduction to Sir Walter Scott, *The Bride of Lammermoor* (Oxford, 1991).

RUFF, JAMES L., Introduction to Lady Caroline Lamb, *Glenarvon*, fac. edn. (New York, 1972).

RYAN, J. S., 'A Possible Australian Source for Miss Havisham', *Australian Literary Studies*, 1 (1963), 134–6.

SANDRIN, ANNY, *Great Expectations* (London, 1988).

SCHMIDGALL, GARY, *Literature as Opera* (New York, 1977).

SCRIVENER, MICHAEL HENRY, *Radical Shelley: The Philosophical Anarchism and Utopian Thought of Percy Bysshe Shelley* (Princeton, 1982).

SEDGWICK, EVE KOSOFSKY, *Tendencies* (London, 1994).

SHIRES, LINDA (ed.), *Rewriting the Victorians: Theory, History, and the Politics of Gender* (London, 1992).

SHOWALTER, ELAINE, 'Desperate Remedies: Sensation Novels of the 1860s', *Victorian Newsletter*, 49 (1976), 1–5.

—— 'Victorian Women and Insanity', *Victorian Studies*, 23 (1980), 157–81.

—— 'Representing Ophelia: Women, Madness, and the Responsibilities of Feminist Criticism', in Patricia Parker and Geoffrey Hartman (eds.), *Shakespeare and the Question of Theory* (London, 1985).

SHUTTLEWORTH, SALLY, ' "The Surveillance of a Sleepless Eye": The Constitution of Neurosis in *Villette*', in George Levine (ed.), *One Culture: Essays in Science and Literature* (Madison, Wis., 1987).

SMITH, LEROY W., *Jane Austen and the Drama of Women* (Totawa, NJ, 1983).

SMITH, OLIVIA, *The Politics of Language 1719–1819* (Oxford, 1984).

SOUTHAM, BRIAN (ed.), *Jane Austen: The Critical Heritage* (London, 1968).

SPACKMAN, BARBARA, *Decadent Genealogies: The Rhetoric of Sickness from Baudelaire to D'Annunzio* (Ithaca, NY, 1989).

SPACKS, PATRICIA, *The Female Imagination: A Literary and Psychological Investigation of Women's Writing* (London, 1976).

—— *Imagining a Self: Autobiography and Novel in Eighteenth-Century England* (Cambridge, Mass., 1976).

SPIVAK, GAYATRI CHAKRAVORTY, 'Three Women's Texts and a Critique of Imperialism', *Critical Inquiry*, 12 (1985), 243–61.

STARES, SUSAN, 'British Seduced Maidens', *Eighteenth-Century Studies*, 14 (1980–1), 109–34.

STONE, HARRY, *Charles Dickens and the Invisible World: Fairy Tales, Fantasy, and Novel Making* (Bloomington, Ind., 1979).

STOREY, GLADYS, *Dickens and Daughter* (London, 1939).

SUMMERS, MONTAGUE, *Essays in Petto* (Edinburgh, 1982).

SUTHERLAND, JOHN, *Victorian Novelists and Publishers* (London, 1976).

—— 'The British Book Trade and the Crash of 1826', in *The Library*, 6th ser., 9 (1987), 148–61.

—— *The Longman Companion to Victorian Fiction* (Harlow, 1988).

—— *Victorian Fiction: Writers, Publishers, Readers* (London, 1995).

SYMONS, JULIAN, Introduction to Wilkie Collins, *The Woman in White* (Harmondsworth, 1974).

TANNER, TONY, *Jane Austen* (London, 1986).

TAYLOR, JENNY BOURNE, *In the Secret Theatre of Home: Wilkie Collins, Sensation Narrative, and Nineteenth-Century Psychology* (London, 1988).

TAYLOR, JOHN TINNON, *Early Opposition to the English Novel: The Popular Reaction from 1760 to 1830* (New York, 1943).

TILLOTSON, KATHLEEN, *Novels of the Eighteen-Forties* (London, 1954).

TODD, JANET, *Sensibility: An Introduction* (London, 1986).

—— *Feminist Literary History: A Defence* (Oxford, 1988).

—— (ed.), *Jane Austen: New Perspectives, Women and Literature*, NS 3 (New York, 1983).

TOMPKINS, J. M. S., *The Popular Novel in England 1770–1800* (London, 1961).

TROMBLEY, STEPHEN, *'All that Summer she was Mad': Virginia Woolf and her Doctors* (London, 1981).

VAEGER, PATRICIA, *Honey-Mad Women: Emancipatory Strategies in Women's Writing* (New York, 1988).

VALOM, MARILYN, *Maternity, Mortality, and the Literature of Madness* (University Park, Penn., 1985).

WACK, MARY F., *Lovesickness in the Middle Ages: The 'Viaticum' and its Commentaries* (Philadelphia, 1990).

WELSH, ALEXANDER, *The Hero of the Waverley Novels* (New Haven, 1963).

WHITE, HENRY ADELBERT, *Sir Walter Scott's Novels on the Stage* (New Haven, 1927).

WILT, JUDITH, *Secret Leaves: The Novels of Sir Walter Scott* (London, 1981).

WILTSHIRE, JOHN, *Jane Austen and the Body* (Cambridge, 1992).

WINNIFRITH, TOM, *A New Life of Charlotte Brontë* (Basingstoke, 1988).

WISE, THOMAS JAMES, and SYMINGTON, JOHN ALEXANDER (eds.), *The*

Brontës: Their Lives, Friendships and Correspondence, 4 vols. (Oxford, 1933).

YATES, EDMUND (ed.), *Celebrities at Home*, 3rd ser. (London, 1879).

Secondary Sources on History and Philosophy of Medicine

ADAMS, PARVEEN, 'Symptoms and Hysteria', *Oxford Literary Review*, 8 (1986), 178–83.

ALAM, CHRIS N., and MERSKEY, H., 'The Development of the Hysterical Personality', *History of Psychiatry*, 3 (1992), 135–65.

ALTSCHULE, M., *Roots of Medical Psychiatry* (New York, 1957).

ANON., *Memoir of George Mann Burrows, M.D., F.L.S., Fellow of the Royal College of Physicians, London (Extracted from the London Medical Gazette of December 11, 1846)* (London, [c.1846]).

ARMSTRONG, DAVID, *The Political Anatomy of the Body: Medical Knowledge in Britain in the Twentieth Century* (Cambridge, 1983).

BEER, GILLIAN, and MARTINS, HERMINIO (eds.), *History of the Human Sciences*, 3/2, Special Issue, 'Rhetoric and Science' (1990).

BULLOUGH, VERN, and BULLOUGH, BONNIE, *Sin, Sickness, and Sanity: A History of Sexual Attitudes* (New York, 1977).

BUSFIELD, JOAN, *Managing Madness: Changing Ideas and Practice* (London, 1986).

BYNUM, WILLIAM F., 'Theory and Practice in British Psychiatry from J. C. Pritchard (1786–1848) to Henry Maudsley (1835–1918)', in T. Ogawa (ed.), *History of Psychiatry* (Osaka, 1982).

——and PORTER, ROY (eds.), *William Hunter and the Eighteenth-Century Medical World* (London, 1985).

——BROWNE, E. J., and PORTER, ROY (eds.), *Dictionary of the History of Science* (Princeton, 1981).

——PORTER, ROY, and SHEPHERD, MICHAEL (eds.), *The Anatomy of Madness: Essays in the History of Psychiatry*, 3 vols. (London, 1985–8).

BYRD, MAX, *Visits to Bedlam* (Columbia, SC, 1974).

——'The Madhouse, the Whorehouse and the Convent', *Partisan Review*, 44 (1977), 268–78.

CAPRA, F., *The Turning Point: Science, Society and the Rising Culture* (New York, 1982).

CARLSON, ERIC T., and DAIN, N., 'The Meaning of Moral Insanity', *Bulletin of the History of Medicine*, 36 (1962), 130–40.

CARSWELL, J., *The South Sea Bubble* (London, 1960).

CASSIRER, ERNST, *The Problem of Knowledge*, trans. William Woglom and Charles Hendel (New Haven, 1950).

CECCIO, JOSEPH, *Medicine in Literature* (London, 1978).

CHAMBERLIN, J. EDWARD, and GILMAN, SANDER L. (eds.), *Degeneration: The Dark Side of Progress* (New York, 1983).

CHESLER, PHYLLIS, *Women and Madness* (New York, 1973).

COHEN, S., and SCULL, A. (eds.), *Social Control and the State* (Oxford, 1983).

COLLIE, MICHAEL, *Henry Maudsley, Victorian Psychiatrist: A Bibliographical Study* (Winchester, 1988).

COOTER, ROGER, *The Cultural Meaning of Popular Science: Phrenology and the Organisation of Consent in Nineteenth Century Britain* (Cambridge, 1984).

DEPORTE, MICHAEL V., *Nightmares and Hobbyhorses: Swift, Sterne and Augustan Ideas of Madness* (San Marino, Calif., 1974).

DESMOND, ADRIAN, *The Politics of Evolution: Morphology, Medicine, and Reform in Radical London* (Chicago, 1989).

DIDI-HUBERMAN, GEORGES, *Invention de l'hystérie: Charcot et l'iconographie photographique de la Salpêtrière* (Paris, 1982).

DIGBY, ANNE, *Madness, Morality and Medicine: A Study of the York Retreat 1796–1914* (Cambridge, 1985).

DODDS, E. R., *The Greeks and the Irrational* (Berkeley, 1951).

DOERNER, KLAUS, *Madmen and the Bourgeoisie: A Social History of Insanity and Psychiatry*, 1969, trans. Joachim Neugroschel and Jean Steinberg (Oxford, 1981).

DONNELLY, MICHAEL, *Managing the Mind: A Study of Medical Psychology in Early Nineteenth-Century Britain* (London, 1983).

EHRENREICH, BARBARA, and ENGLISH, DEIRDRE, *Complaints and Disorders: The Sexual Politics of Sickness* (New York, 1973).

FISSELL, MARY E., *Patients, Power, and the Poor in Eighteenth-Century Bristol* (Cambridge, 1991).

FOUCAULT, MICHEL, *Madness and Civilisation: A History of Insanity in the Age of Reason*, trans. and abridged by Richard Howard (London, 1967).

——*The Order of Things: An Archaeology of the Human Sciences* (1966), trans. anon. (New York, 1970).

——*The Birth of the Clinic*, trans. A. M. Sheridan (London, 1972).

——*Histoire de la folie à l'âge classique*, 2e edn. suivie de 'Mon corps, ce papier, ce feu' (Paris, 1972).

——*Discipline and Punish: The Birth of the Prison*, trans. A. M. Sheridan (London, 1977).

GALLAGHER, CATHERINE, and LAQUEUR, THOMAS (eds.), *The Making of the Modern Body: Sexuality and Society in the Nineteenth Century* (Berkeley, 1987).

GALLINI, CLARA, *La sonnambula meravigliosa: magnetismo e ipnotismo nell'ottocento italiano* (Milan, 1983).

GEISON, G. L., *Michael Foster and the Cambridge School of Physiology: The Scientific Enterprise in Late Victorian Society* (Princeton, 1978).

GILMAN, SANDER L., *Seeing the Insane: A Cultural History of Madness and Art in the Western World* (New York, 1982).

——*Difference and Pathology: Stereotypes of Sexuality, Race and Madness* (New York, 1985).

——'The Madman as Artist: Medicine, History and Degenerate Art', in *Journal of Contemporary History*, 20 (1985), 575–97.

——*Disease and Representation: Images of Illness from Madness to AIDS* (Ithaca, NY, 1988).

——(ed.), *The Faces of Madness: Hugh W. Diamond and the Origin of Psychiatric Photography* (New York, 1976).

GOFFMAN, ERVING, *Asylums: Essays on the Social Situation of Asylum Patients and Other Inmates* (New York, 1961).

GORDON, FELICIA, 'Psychiatric Monomaniacs', *Gender and History*, 1 (1989), 219–25.

HARE, E. H., 'Was Insanity on the Increase?', *British Journal of Psychiatry*, 142 (1983), 439–55.

HEATH, STEPHEN, *The Sexual Fix* (Harmondsworth, 1982).

HERZLICH, CLAUDINE, and PIERRET, JANINE, *Illness and Self in Society* (1984), trans. Elborg Forster (Baltimore, 1987).

HOWELLS, J. G. (ed.), *World History of Psychiatry* (New York, 1968).

HUNTER, K. M., *Doctors' Stories: The Narrative Structure of Medical Knowledge* (Princeton, 1991).

HUNTER, RICHARD, and MACALPINE, IDA, *Three Hundred Years of Psychiatry, 1535–1860* (London, 1963).

——'Dickens and Conolly: An Embarrassed Editor's Disclaimer', *Times Literary Supplement*, 11 Aug. 1961, 534–5.

ILLICH, IVAN, *et al.*, *Disabling Professions* (London, 1977).

INGLEBY, DAVID, *Critical Psychiatry: The Politics of Mental Health* (Harmondsworth, 1981).

——'The Social Construction of Mental Illness', in P. Wright and A. Treacher (eds.), *The Problem of Medical Knowledge* (Edinburgh, 1982).

JACKSON, M. W., *Goethe's Law and Order: Art and Nature in 'Elective Affinities'*, doctoral dissertation, University of Cambridge, 1991.

JEWSON, N. 'The Disappearance of the Sick Man from Medical Cosmology 1770–1870', *Sociology*, 10 (1976), 225–44.

JONES, KATHLEEN, *Lunacy, Law and Conscience 1744–1845* (London, 1960).

JORDANOVA, LUDMILLA, *Sexual Visions: Images of Gender in Science and Medicine between the Eighteenth and Twentieth Centuries* (New York, 1989).

KELLER, EVELYN FOX, *Reflections on Gender and Science* (New Haven, 1984).

KING, HELEN, 'From Parthenos to Gyne: The Dynamics of Category,' doctoral dissertation, University College, London, 1985.

——'Once upon a Text: The Hippocratic Origins of Hysteria', in George Rousseau and Roy Porter (eds.), *Hysteria in Western Civilization* (Los Angeles, forthcoming).

KNIBIEHLER, YVONNE, and FOUQUET, CATHERINE, *La Femme et les médecins: analyse historique* (Paris, 1983).

KUHN, THOMAS S., *The Structure of Scientific Revolutions* (Chicago, 1970).

LAING, R. D., *The Divided Self: A Study of Sanity and Madness* (London, 1960).

——and ESTERTON, A., *Sanity, Madness and the Family* (New York, 1964).

LAQUEUR, THOMAS, *Making Sex: Body and Gender from the Greeks to Freud* (Cambridge, Mass., 1990).

LAWRENCE, CHRISTOPHER, 'The Nervous System and Society in the Scottish Enlightenment', in Barry Barnes and Steve Shapin (eds.), *Natural Order: Historical Studies of Scientific Culture* (London, 1979).

——'Incommunicable Knowledge: Science, Technology and the Clinical Art in Britain 1850–1914', *Journal of Contemporary History*, 20 (1985), 503–20.

MACALPINE, IDA, and HUNTER, RICHARD, *George III and the Mad-Business* (London, 1969).

MACDONALD, MICHAEL, *Mystical Bedlam: Madness, Anxiety and Healing in Seventeenth Century England* (Cambridge, 1981).

MASSON, JEFFREY M., *A Dark Science: Women, Sexuality, and Psychiatry in the Nineteenth Century* (New York, 1986).

MERSKEY, HAROLD, and POTTER, PAUL, 'The Womb Lay Still in Ancient Egypt', *British Journal of Psychiatry*, 154 (1989), 751–3.

MICALE, MARK S., 'Hysteria and its Historiography: The Future Perspective', *History of Psychiatry*, 1 (1990), 32–124.

——*Approaching Hysteria: Disease and its Interpretations* (Princeton, 1995).

——and PORTER, ROY (eds.), *Discovering the History of Psychiatry* (Oxford, 1994).

MOSCUCCI, ORNELLA, *The Science of Woman: Gynaecology and Gender in England 1800–1929* (Cambridge, 1990).

MOSEDALE, SUSAN SLEETH, 'Science Corrupted: Victorian Biologists Consider "The Woman Question"', *Journal of the History of Biology*, 11 (1978), 1–55.

OPPENHEIM, JANET, *'Shattered Nerves': Doctors, Patients, and Depression in Victorian England* (New York, 1991).

OWEN, ALEX, *The Darkened Room: Women, Power, and Spiritualism in Late Nineteenth-Century England* (London, 1989).

PARADIS, JAMES, and POSTLEWAIT, THOMAS (eds.), *Victorian Science and Victorian Values: Literary Perspectives* (New York, 1981).

PARRY-JONES, WILLIAM LLEWELLYN, *The Trade in Lunacy: A Study of Private Madhouses in England in the Eighteenth and Nineteenth Centuries* (London, 1971).

PETERSON, DALE (ed.), *A Mad People's History of Madness* (Pittsburgh, 1982).

PETERSON, M. JEANNE, *The Medical Profession in Mid-Victorian London* (Berkeley, 1978).

PICK, DANIEL, *Faces of Degeneration: A European Disorder c.1848–c.1918* (London, 1989).

PORTER, ROY, 'Shutting People Up', *Social Studies of Science*, 12 (1982), 467–76.

——'The Rage of Party: A Glorious History of Psychiatry?', *Medical History*, 27 (1983), 35–50.

——*A Social History of Madness: Stories of the Insane* (London, 1987).

——*Mind Forg'd Manacles: A History of Madness in England from the Restoration to the Regency* (London, 1987).

——(ed.), *Patients and Practitioners: Lay Perceptions of Medicine in Pre-industrial Society* (Cambridge, 1985).

——and PORTER, DOROTHY, *Patient's Progress: Doctors and Doctoring in Eighteenth-Century England* (Oxford, 1989).

RIPA, YANNICK, *Women and Madness: The Incarceration of Women in Nineteenth-Century France*, trans. Catherine du Peloux Mangé (Cambridge, 1990).

RISSE, GUENTER B., 'Hysteria at the Edinburgh Infirmary: The Construction and Treatment of a Disease 1770–1800', *Medical History*, 32 (1988), 1–22.

ROBERTS, MARIE MULVEY, and PORTER, ROY (eds.), *Literature and Medicine during the Eighteenth Century* (London, 1993).

ROBINS, JOSEPH, *Fools and Mad: A History of the Insane in Ireland* (Dublin, 1986).

ROSEN, GEORGE, *Madness in Society: Chapters in the Historical Sociology of Mental Illness* (London, 1968).

ROTH, MARTIN, and KROLL, JEROME, *The Reality of Mental Illness* (Cambridge, 1986).

ROTHFIELD, LAWRENCE, *Vital Signs: Medical Realism in Nineteenth-Century Fiction* (Princeton, 1992).

ROTHMAN, DAVID, *The Discovery of the Asylum* (Boston, 1971).

ROUSSEAU, G. S., 'Science and Literature, the State of the Art', *Isis*, 69 (1978), 583–91.

——'Literature and Medicine: The State of the Field', *Isis*, 72 (1981), 406–24.

——*Enlightenment Borders: Pre- and Postmodern Discourses: Medical, Scientific* (Manchester, 1991).

——*Enlightenment Crossings: Pre- and Postmodern Discourses: Anthropological* (Manchester, 1991).

——(ed.), *The Languages of Psyche: Mind and Body in Enlightenment Thought* (Berkeley, 1991).

——and PORTER, ROY (eds.), *The Ferment of Knowledge: Studies in the Historiography of Eighteenth-Century Science* (Cambridge, 1980).

SCARRY, ELAINE, *The Body in Pain* (New York, 1985).

SCULL, ANDREW, 'From Madness to Mental Illness: Medical Men as Moral Entrepreneurs', *European Journal of Sociology*, 16 (1975), 218–51.

——*Museums of Madness: The Social Organisation of Insanity in Nineteenth-Century England* (London, 1979).

——'A Convenient Place to Get Rid of Inconvenient People: The Victorian Lunatic Asylum', in A. D. King (ed.), *Buildings and Society* (London, 1980).

——(ed.), *Madhouses, Mad-Doctors, and Madmen: The Social History of Psychiatry in the Victorian Era* (London, 1981).

——'A Brilliant Career? John Conolly and Victorian Psychiatry', *Victorian Studies*, 27 (1984), 203–35.

——*Decarceration: Community Treatment and the Deviant—A Radical View* (Cambridge, 1984).

——*The Most Solitary of Afflictions: Madness and Society in Britain 1700–1900* (London, 1993).

SEDGWICK, PETER, *Psycho Politics* (London, 1982).

SHORTT, S. E. D., *Victorian Lunacy: Richard M. Bucke and the Practice of Late Nineteenth Century Psychiatry* (Cambridge, 1986).

SHOWALTER, ELAINE, *The Female Malady: Women, Madness, and English Culture 1830–1980* (London, 1987).

SKULTANS, VIEDA, *Madness and Morals: Ideas on Insanity in the Nineteenth Century* (London, 1975).

——*English Madness: Ideas on Insanity 1580–1890* (London, 1979).

SMALL, HELEN, Review of Allan Ingram, *The Madhouse of Language*, *History of Psychiatry*, 4 (1993), 456–8.

——' "In the Guise of Science": Literature and the Rhetoric of Nineteenth-Century English Psychiatry', *History of the Human Sciences*, 7 (1994), 27–55.

SMITH, ROGER, *Trial by Medicine: Insanity and Responsibility in Victorian Trials* (Edinburgh, 1981).

SONTAG, SUSAN, *Illness as Metaphor* (Harmondsworth, 1977).

STEPAN, NANCY, *The Idea of Race in Science: Great Britain 1800–1960* (London, 1982).

STILL, ARTHUR, and VELODY, IRVING (eds.), *Rewriting the History of Madness: Studies in Foucault's 'Histoire de la folie'* (London, 1992).

Szasz, Thomas S. (ed.), *The Myth of Mental Illness: Foundations of a Theory of Personal Conduct* (London, 1961).

——*The Manufacture of Madness: A Comparative Study of the Inquisition and the Mental Health Movement* (London, 1971).

——*The Age of Madness: The History of Involuntary Mental Hospitalization Presented in Selected Texts* (London, 1975).

Taylor, Barbara, *Eve and the New Jerusalem: Socialism and Feminism in the Nineteenth Century* (London, 1983).

Ussher, Jane M., *Women's Madness: Misogyny or Mental Illness?* (London, 1991).

Veith, Ilza, *Hysteria: The History of a Disease* (Chicago, 1945).

Waddington, Ivan, *The Medical Profession and the Industrial Revolution* (Dublin, 1984).

Walker, Nigel, *Crime and Insanity in England*, 2 vols. (Edinburgh, 1968).

Walton, John K., 'Lunacy in the Industrial Revolution: A Study of Asylum Admissions in Lancashire 1848–50', *Journal of Social History*, 13 (1979), 1–22.

Williams, Katherine E., 'Hysteria in Seventeenth-Century Case Records and Unpublished Manuscripts', *History of Psychiatry*, 1 (1990), 383–401.

Winter, Alison, ' "The Island of Mesmeria": The Politics of Mesmerism in Early Victorian Britain', doctoral dissertation, University of Cambridge, 1992.

Yeazell, Ruth, and Neil Hertz (eds.), *Sex, Politics, and Science in the Nineteenth-Century Novel: Essays from the English Institute* (Baltimore, 1986).

Young, Robert M., *Mind, Brain and Adaption in the Nineteenth Century: Cerebral Localization and its Biological Context from Gall to Ferrier* (Oxford, 1970).

Youngson, A. J., *The Scientific Revolution in Victorian Medicine* (London, 1976).

Zainaldin, Jamil S., and Tyor, Peter L., 'Asylum and Society: An Approach to Industrial Change', *Journal of Social History*, 13 (1979), 23–48.

Zilboorg, Gregory, *A History of Medical Psychology* (New York, 1941).

Critical Theory

Bernheimer, Charles, and Kahane, Claire (eds.), *In Dora's Case: Freud—Hysteria—Feminism* (London, 1985).

CIXOUS, HÉLÈNE, 'Tancredi Continues', in Susan Sellers (ed.), *Writing Differences: Readings from the Seminar of Hélène Cixous* (Milton Keynes, 1988).

——and CLÉMENT, CATHERINE, *La Jeune Née* (1975); trans. as *The Newly Born Woman* by Betsy Wing (Minneapolis, 1986).

CLÉMENT, CATHERINE, *L'Opéra; ou, La Défaite des femmes* (1979); trans. as *Opera; or, The Undoing of Women* by Betsy Wing (London, 1989).

CULLER, JONATHAN, *On Deconstruction: Theory and Criticism after Structuralism* (Ithaca, NY, 1982; repr. London, 1983).

DELEUZE, GILLES, and GUATTARI, FÉLIX, *L'Anti-Oedipe: Capitalisme et schizophrénie* (Paris, 1972).

——*Milles Plateaux* (Paris, 1980).

——*Semiotext(e)* (New York, 1983).

DERRIDA, JACQUES, 'Cogito et histoire de la folie', in Derrida, *L'Écriture et la différence* (Paris, 1967); trans. as 'Cogito and the History of Madness' by Alan Bass, in Derrida, *Writing and Difference* (London, 1978).

FELMAN, SHOSHANA, *La Folie et la chose littéraire* (1978); trans. as *Writing and Madness (Literature/Philosophy/Psychoanalysis)* by Martha Noel Evans with the author (Ithaca, NY, 1985).

——*Literature and Psychoanalysis: The Question of Reading: Otherwise* (New Haven, 1981).

——'Women and Madness: The Critical Phallacy', *Diacritics*, 5/4 (1975), 2–10.

FELSKI, RITA, *Beyond Feminist Aesthetics: Feminist Literature and Social Change* (London, 1989).

GALLAGHER, CATHERINE, 'The Networking Muse: Aphra Behn as Heroine of Frankness and Self-Discovery', *Times Literary Supplement*, 10 Sept. 1993, 3–4.

GALLOP, JANE, *Feminism and Psychoanalysis: The Daughter's Seduction* (Basingstoke, 1982).

——*Thinking through the Body* (New York, 1988).

GENETTE, GÉRARD, 'Structuralism and Literary Discourse', in Genette, *Figures of Literary Discourse*, trans. A. M. Sheridan (Columbia, SC, 1982).

HOMANS, MARGARET, 'Feminist Fictions and Feminist Theories of Narrative', *Narrative*, 2 (1994), 3–16.

HUNTER, DIANNE, 'Hysteria, Psychoanalysis and Feminism: The Case of Anna O——', *Feminist Studies*, 9 (1993), 3–16.

IRIGARAY, LUCE, *Le Langage des déments* (Paris, 1973).

——*Ce sexe qui n'en est pas un* (1977); trans. as *This Sex which is not One* by Catherine Porter with Caroline Burke (Ithaca, NY, 1985).

——*Parler n'est jamais neutre* (Paris, 1985).

JACOBUS, MARY (ed.), *Women Writing and Writing about Women* (London, 1979).

——Review of *The Madwoman in the Attic* and *Shakespeare's Sisters*, *Signs*, 6 (1981), 517–23.

——*Reading Woman: Essays in Feminist Criticism* (London, 1986).

JARDINE, ALICE, *Gynesis: Configurations of Woman and Modernity* (Ithaca, NY, 1985).

JONES, ANN ROSALIND, 'Julia Kristeva on Femininity: The Limits of a Semiotic Politics', *Feminist Review*, 18 (1984), 56–73.

JOUVE, NICOLE WARD, *White Woman Speaks with Forked Tongue: Criticism as Autobiography* (London, 1991).

KAPLAN, CORA, *Sea Changes: Culture and Feminism* (London, 1986).

KRISTEVA, JULIA, *Semeiotikè: Recherches pour une sémanalyse* (Paris, 1969).

——*Des chinoises* (1974); trans. as *About Chinese Women* by Anita Burrows (London, 1977).

——*La Révolution du langage poétique* (Paris, 1974).

——*Pouvoirs de l'horreur: Essai sur l'abjection* (Paris, 1980).

——*Histoires d'amour* (Paris, 1983).

——*Desire in Language*, ed. Leon S. Roudiez, trans. Thomas Gora, Alice Jardine, and L. Roudiez (Oxford, 1984).

MARCUS, JANE, 'The Asylums of Antaeus: Women, War, and Madness—Is there a Feminist Fetishism?', in H. Aram Veeser (ed.), *The New Historicism* (London, 1990).

MARKS, ELAINE, and DE COURTIVRON, ISABELLE (eds.), *New French Feminisms: An Anthology* (Brighton, 1981).

MITCHELL, JULIET, *Women: The Longest Revolution: Essays on Feminism, Literature and Psychoanalysis* (London, 1984).

ROSE, JACQUELINE, *Sexuality in the Field of Vision* (London, 1986).

SEDGWICK, EVE KOSOFSKY, *Between Men: English Literature and Male Homosocial Desire* (New York, 1985).

SHOWALTER, ELAINE, 'On Hysterical Narrative', *Narrative*, 1 (1993), 24–35.

Other Sources

ABRAY, JANE, 'Feminism in the French Revolution', *American Historical Review*, 80 (1975), 43–62.

BARBER, PAUL, *Vampires, Burial, and Death: Folklore and Reality* (New Haven, 1988).

BINDMAN, DAVID (ed.), *The Shadow of the Guillotine: Britain and the French Revolution (1789–1820)* (New Haven, 1983).

BOTT, CATHERINE, *Mad Songs*, CD no. 433 187-2 (Decca–L'Oiseau Lyre; London, 1993).

BURKE, EDMUND, *Reflections on the Revolution in France and on the Proceedings in Certain Societies in London Relative to that Event* (1790), ed. L. G. Mitchell (Oxford, 1993).

COTT, NANCY, 'Passionlessness: An Interpretation of Victorian Sexual Ideology 1790–1850', *Signs*, 4 (1978), 219–36.

DAVIDOFF, LEONORE, and HALL, CATHERINE, *Family Fortunes: Men and Women of the English Middle Class 1780–1850* (London, 1987).

DOUGLAS, MARY, *Purity and Danger: An Analysis of Concepts of Pollution and Taboo* (London, 1966).

——*Natural Symbols: Explorations in Cosmology* (London, 1970).

ELLIOTT, MARIANNE, *Partners in Revolution: The United Irishmen and France* (New Haven, 1982).

FFOULKES, MRS, *The Story of Sarah Fletcher* (n.p., n.d.).

FORBES, THOMAS ROGERS, 'Crowner's Quest', *Transactions of the American Philosophical Society*, 68/1 (1978).

The Girlhood of Queen Victoria: A Selection from Her Majesty's Diaries between the Years 1832 and 1840. Published by the Authority of His Majesty the King, ed. Viscount Esher, 2 vols. (London, 1912).

HANLEY, KEITH, and SELDEN, RAMAN (eds.), *Revolution and English Romanticism: Politics and Rhetoric* (Hemel Hempstead, 1990).

HORNER, FRANCIS, *The Horner Papers: Selections from the Letters and Miscellaneous Writings of Francis Horner, M.P. 1795–1817*, ed. Kenneth Bourne and William Banks Taylor (Edinburgh, 1994).

HUNT, LEIGH, MS Letter to Bryan Waller Procter, 22–3 June 1857, University of Iowa Library (NP 0596785).

LANDES, JOAN B., *Women and the Public Sphere in the Age of the French Revolution* (Ithaca, NY, 1988).

LOUGH, JOHN, and MERSON, ELIZABETH, *John Graham Lough 1788–1876: A Northampton Sculptor* (Woodbridge, 1987).

MACDONALD, MICHAEL, and MURPHY, TERENCE R., *Sleepless Souls: Suicide in Early Modern England* (Oxford, 1990).

MAYHEW, HENRY, *London Labour and the London Poor*, 2 vols. (London, 1851).

OUTRAM, DORINDA, *The Body and the French Revolution: Sex, Class, and Political Culture* (New Haven, 1989).

ROSA, ANNETTE, *Citoyennes: Les Femmes et la Révolution Française* (Paris, 1988).

SEWELL, WILLIAM H., JR., 'Le Citoyen/la citoyenne: Activity, Passivity and the Revolutionary Concept of Citizenship', in *The French Revolution and the Creation of Modern Political Culture*, ii: *The Political Culture of the French Revolution*, ed. C. Lucas (Oxford, 1988).

SPARROW, JOHN, *Grave Epigrams* (Bembridge, 1974).

—— *Lapidaria*, 8 vols. (Cambridge, 1943–81).

STEDMAN, EDITH, Postscript to Mrs Ffoulkes, *The Story of Sarah Fletcher*, MS, private collection.

STOCKING, GEORGE W., JR., *Victorian Anthropology* (New York, 1987).

STONE, LAWRENCE, *The Family, Sex and Marriage in England 1500–1800* (London, 1977).

STUART, DOROTHY MARGARET, *The Daughters of George III* (London, 1939).

SULEIMAN, SUSAN, *The Female Body in Western Culture* (Cambridge, Mass., 1986).

TRUDGILL, ERIC, *Madonnas and Magdalens: The Origins and Development of Victorian Sexual Attitudes* (London, 1976).

UMBREVILLE, E., *Lex Coronatoria* (London, 1761).

VICINUS, MARTHA, *Suffer and be Still: Women in the Victorian Age* (Bloomington, Ind., 1973).

VICKERY, AMANDA, 'Golden Age to Separate Spheres? A Review of the Categories and Chronology of English Women's History', *Historical Journal*, 36 (1993), 383–414.

WAINEWRIGHT, THOMAS GRIFFITHS, *Essays and Criticisms by Thomas Griffiths Wainewright*, ed. W. Carew Hazlitt (London, 1880).

WALPOLE, HORACE, *The Letters of Horace Walpole, Fourth Earl of Orford*, 16 vols., ed. Mrs Paget Toynbee (Oxford, 1905).

WATSON, J. R. (ed.), *The French Revolution in English Literature and Art: The Yearbook of English Studies*, 19 (London, 1989).

Index